Canto is an imprint offering a range of
titles, classic and more recent, across a
broad spectrum of subject areas and
interests. History, literature, biography,
archaeology, politics, religion, psychology,
philosophy and science are all represented
in Canto's specially selected list of titles,
which now offers some of the best and
most accessible of Cambridge publishing to
a wider readership.

THE THEORY OF EVOLUTION

THE THEORY OF EVOLUTION

JOHN MAYNARD SMITH

Published by the Press Syndicate of the University of Cambridge
The Pitt Building, Trumpington Street, Cambridge CB2 1RP
40 West 20th Street, New York, NY 10011–4211 USA
10 Stamford Road, Oakleigh, Melbourne 3166, Australia

First published by Penguin Books 1958
Second edition 1966
Third edition 1975
Canto edition first published by Cambridge University Press 1993
Reprinted 1995, 1997

Printed in Great Britain at the University Press, Cambridge

A catalogue record for this book is available from the British Library

ISBN 0 521 45128 0 paperback

Contents

Figures

Foreword to the Canto Edition

This book, in its original shorter edition, was my first introduction to John Maynard Smith and one of my first introductions to evolution. I bought it as a schoolboy, instantly captivated by the jacket blurb and author's photograph. The wild, nutty-professor hair, aslant like the pipe in the cheerfully smiling mouth; even the obviously intelligent eyes seemed somehow askew as they laughed their way through thick, round glasses (this was before John Lennon made them fashionable) badly in need of a clean. The picture perfectly complemented the quirky biographical note: 'Deciding that aeroplanes were noisy and old-fashioned, he entered University College London, to study zoology.' I kept peeping at the back cover as I read, then returned to the text with a smile and renewed confidence that this was a man whose views I wanted to hear. I have known him personally now for twenty-six years and my initial impression has only deepened. This is a man whose views I want to hear, and so says everyone who knows him or reads his books, or even casually encounters him. At a conference for example.

Readers of 'campus novels' know that a conference is where you can catch academics at their worst. The conference bar, in particular, is the academy in microcosm. Professors huddle together in exclusive, conspiratorial corners, talking not about science or scholarship but about 'tenure-track hiring' (their word for jobs) and 'funding' (their word for money). If they do talk shop, too often it will be to make an impression rather than to enlighten. John Maynard Smith is a splendid, triumphant, lovable exception. He values creative ideas above money, plain language above jargon. He is always the centre of a lively, laughing crowd of students and young research workers of both sexes. Never mind the lectures or the 'workshops'; be blowed to the motor coach excursions to local beauty spots; forget your fancy visual aids and radio microphones; the only thing that really matters at a conference is that John Maynard Smith must be in residence and there must be a spacious, convivial bar. If he can't manage the dates you have in mind, you must just reschedule the conference. He doesn't have to give a formal talk (although he is a riveting speaker) and he doesn't have to chair a formal session (although he is a wise, sympathetic and witty chairman). He has only to turn up and your conference will succeed. He will charm and amuse the young research workers, listen to their stories, inspire them, rekindle enthusiasms that might be flagging, and send them back to their laboratories or their muddy fields, enlivened and invigorated, eager to try out the new ideas he has generously shared with them.

Not just ideas but knowledge, too. He sometimes quaintly poses as a workaday engineer who doesn't know anything about animals and plants. He *was* originally trained as an engineer, and the mathematical outlook and skills of his old vocation invigorate his present one. But he has been a professional biologist for a good forty years and a naturalist since childhood. He is leagues away from that familiar menace: the brash physical scientist who thinks he can wade in and clean up biology because, no matter how poorly he shows up against his fellow physicists, he at least knows more mathematics than the average biologist. John does know more mathematics, more physics and more engineering than the average biologist. But he also knows more biology than the average biologist. And he is incomparably more gifted in the arts of clear thinking and communicating than most physicists or biologists or anybody else. More, like a finely tuned antenna, he has the rare gift of biological intuition. Walk through wild country with him as I am privileged to have done, and you learn not just facts about natural history but the right way to ask questions about those facts. Better still, unlike some theorists, he has deep respect for good naturalists and experimentalists, even if they lack his own theoretical clout. He and I were once being shown around the Panama jungle by a young man, one of the staff of the Smithsonian tropical research station, and John whispered to me: 'What a privilege to listen to a man who really loves his animals.' I agreed, though the young man in this case was a forester and his 'animals' were various species of palm tree.

He is generous and tolerant of the young and aspiring, but a merciless adversary when he detects a dominating, powerful academic figure in pomposity or imposture. I have seen him turn red with anger when confronted with a piece of rhetorical duplicity from a senior scientist before a young audience. If you ask him to name his own greatest virtue I suspect that, though he would be modest about nearly all his many skills and accomplishments, he would make one claim for himself: that he cares passionately about the truth. He is one of the few opponents who is seriously feared by creationist debaters. The slickest of these, like glib lawyers paid to advocate a poor case, are accustomed to bamboozling innocent audiences. They are eager to take on respectable scientists in debate, partly because they gain kudos and credibility from sharing a platform, on apparently equal terms, with a legitimate scholar. But they fear John Maynard Smith because, though he doesn't enjoy it, he always trounces them. Only a few weeks ago an anti-evolutionist author, basking in the short-term publicity that grows out of publishers' buying journalists lunch, was booked to have a debate in Oxford. Press and television interest had been easily whipped up, and the author's publishers must have been rubbing their hands with glee. Then the unfortunate fellow discovered who his opponent was to be: John Maynard Smith! He instantly backed out, and his supporters could do nothing to change his mind. If the debate had taken place John would indeed have routed him. But he'd have done it without rancour, and afterwards he'd have bought the wretched man a drink and even got him laughing.

I suppose some successful scientists make their careers by hammering away at one experimental technique that they are good at, and by gathering a gang of co-workers to do the donkey work. Their continued success rests primarily on their ability to coax a steady supply of money out of the government. John Maynard Smith, by contrast, makes his way almost entirely by original thought, needing to spend very little money, and there is scarcely a branch of evolutionary or population genetic theory that has not been illuminated by his vivid and versatile inventiveness. He is one of that rare company of scientists that changes the way people think. Together with only a handful of others, including W. D. Hamilton and G. C. Williams, Maynard Smith is one of today's leading Darwinians. Perennially versatile, he has also made important contributions to the theory of biomechanics, of ecology, and of animal behaviour, in which he was largely responsible for promoting the persistently fashionable methods of Game Theory. He is in the forefront of the study of sex, probably the most baffling topic in modern evolutionary theory. Indeed he was largely responsible for recognizing that sex constituted a problem in the first place, the problem now universally known by his phrase, 'the twofold cost of sex'.

He is an infectiously felicitous phrasemaker. His coinings have become a prevailing shorthand among the cognoscenti – 'Genetic Hitch-hiking', 'the Sir Philip Sidney Game', 'chaps' as an abbreviation for *Homo sapiens*, 'Partridge's Fallacy', the 'Haystack Model' – I won't spell out their meanings in detail, but you could fill a small dictionary with words and phrases that he introduced and which are now understood and daily used by evolutionary biologists the world over. He is also responsible for reviving and promulgating the earlier coinings of his mentor, the formidable J. B. S. Haldane: 'Pangloss's Theorem', 'The Bellman's Theorem' (What I tell you three times is true) and 'Aunt Jobiska's Theorem' (It's a fact the whole world knows). In turn, new generations of biologists are inspired to create their own Maynard Smithian phrases – 'the Beau Geste Effect', 'the Vicar of Bray Theory' – to lighten and refresh the pages of normally staid and rather dull academic journals. The pompous high priests of 'political correctness' don't like this kind of verbal informality. Maynard Smith, like Haldane before him, is too big a man to go along with their puritanical emasculation of language (and if my use of 'emasculation' gives offence to somebody, what a pity).

The qualities that make John Maynard Smith the life and soul of a good conference, the nemesis of creationists and charlatans, and the inspiration of so much youthful research, are also the qualities that make him the ideal author of a book for intelligent, critical laypeople. This book which, thanks to Cambridge University Press, he will now have to call something other than 'My little Penguin', never had the flavour of ephemerality. Publishers never needed to buy lunches in order to get this book noticed. Through three editions and numerous reprintings, it has simply won its own place on the shelves of students and the generally literate; a staple that has seen silly fads and frothy fancies come and go. Few people in the world are better qualified than John Maynard Smith to explain evolution to us, and no subject more than evolution deserves such a talented teacher. You can hear his clear, logical, patient tones on every page. Not least, there is a total absence of pretentious languaging-up. Like Darwin himself, Maynard Smith knows that his story is intrinsically interesting enough and important enough to need no more than clear, patient, honest exposition.

It is a measure both of the brilliance of the book and the endurance of the neo-Darwinian synthesis itself that the 1975 text can stand its ground without revision today. There have, of course, been exciting new developments in the field. It would be worrying if there had not, and they are discussed in his new Introduction. But the fundamental ideas and the great bulk of the detailed assertions of the original book remain as important and as true as ever. The new Introduction itself is an elegant essay which can be recommended in its own right as a summary of important recent developments in evolutionary theory.

Darwin's theory of evolution by natural selection is the only workable explanation that has ever been proposed for the remarkable fact of our own existence, indeed the existence of all life wherever it may turn up in the universe. It is the only known explanation for the rich diversity of animals, plants, fungi and bacteria; not just the leopards, kangaroos, Komodo dragons, dragonflies, Corncrakes, Coast redwood trees, whales, bats, albatrosses, mushrooms and bacilli that share our time, but the countless others – tyrannosaurs, ichthyosaurs, pterodactyls, armour-plated fishes, trilobites and giant sea scorpions – that we know only from fossils but which, in their own aeons, filled every cranny of the land and sea. Natural selection is the only workable explanation for the beautiful and compelling illusion of 'design' that pervades every living body and every organ. Knowledge of evolution may not be strictly useful in everyday commerce. You can live some sort of life and die without ever hearing the name of Darwin. But if, before you die, you want to understand why you lived in the first place, Darwinism is the one subject that you must study. This book is the best general introduction to the subject now available.

RICHARD DAWKINS

Preface

The main unifying idea in biology is Darwin's theory of evolution through natural selection. This idea was put forward when the study of fossils was still in its infancy, so that little was known of the actual course of organic evolution; when few studies had been made of the geographical variation of animals and plants; and, most important of all, before the mechanism of inheritance was understood. Yet recent advances in these various fields have confirmed Darwin's ideas, and in fact can only be fitted together to tell a coherent story if the theory of natural selection is accepted.

In this book I have tried to explain the causes of evolution in terms of the processes of variation, selection, and inheritance which can be seen to occur among living animals and plants. Work in many different fields of biology is relevant to such an inquiry. Unhappily no one biologist can hope to be familiar with all these fields, and I am no exception to this rule. Yet it seemed worth while to attempt a book covering the whole subject of modern evolution theory. Many books have been written, mainly for specialists, covering particular aspects of the subject; some of these are listed in the bibliography. I hope the present book will be of value to the non-specialist in summarizing a set of ideas which, taken together, form perhaps the most important contribution yet made by biologists to our understanding of what the world is like and how it came to be like that.

That I have been able to make the attempt at all I owe mainly to those who were first my teachers and are now my colleagues at University College London. As an undergraduate

I had the good fortune to attend Professor D. M. S. Watson's course in vertebrate anatomy, and learnt that the study of fossils, which in lesser hands can so easily degenerate into a tale of prehistoric wonders, can be made to reveal the way of life of animals now extinct and to show why evolution took the course it did.

In those sections of the book dealing with embryology and genetics, and particularly with the relations of these two subjects to each other, I owe a great deal to discussions with Dr D. R. Newth. To him, and to the many other colleagues whom I have plagued with questions I am most grateful.

But my greatest debt is to Dr Helen Spurway and to Professor J. B. S. Haldane, in whose department most of my research has been carried out. I have referred directly to some of their published work, but there are few subjects discussed in this book which I have not at one time or another also discussed with them, and from such discussions I have taken many of the ideas and examples quoted. If I have borrowed too many ideas from them without acknowledgement, I hope they will forgive me; in any case I feel certain that they are both sufficiently fertile of new ideas not to notice the loss.

The chapter dealing with palaeontology has been read in manuscript by Miss Pamela Lamplugh Robinson, and those dealing with speciation by Mr R. B. Freeman. The main part of the manuscript has been read by Professor Haldane and by Dr Newth. I thank them all for pointing out a number of errors, and for making many helpful suggestions. The mistakes which remain are my own. My thanks are also due to Mr W. B. Mackie for the care he has taken in preparing the diagrams.

Finally, a word to the reader about the principles I have tried to follow in writing this book. Before starting a formal training in biology, I had read a number of books about evolution, some intended primarily for specialists and some for laymen. Although there were always, in the former kind, passages which I could not follow, I found such books more satisfying than those written for laymen, since in the latter I had always the feeling that difficulties were being slurred over. I have tried to avoid this fault. Although I have not assumed any specialized knowledge in the reader, and when possible have drawn my examples from familiar animals and plants, I have not omitted any subjects merely because they are difficult.

I am aware that there is therefore a risk that some parts of this book will prove rather hard going. This is most likely to be so in those sections which discuss the genetic aspects of evolution. People seem to be divided rather sharply into those who find Mendelian genetics easy, and those who find it incomprehensible. For the benefit of the latter, I have tried to concentrate the more difficult genetic arguments into a few chapters, which can be skipped by those with no taste for this kind of argument. But I hope that not too many readers will find this necessary.

Preface to the Second Edition

In the eight years since this book was written, much work has
been done on artificial selection, on natural selection in wild
populations, on speciation, and on palaeontology. But although
many details have been filled in, this work has not altered the
general picture presented in the first edition. The need for a
second edition arises because of advances in molecular genetics.
The critical discovery – the structure of DNA – was published
in 1953. But it has taken some ten years to see the relevance of
this discovery for evolution theory.

In effect, only two theories of evolution have ever been put
forward; one, originating with Lamarck, suggests that the
evolutionary origin of adaptations lies in the adaptation of
individuals during development and the hereditary transmission of these acquired adaptations; the other, originating
with Darwin and formulated in its 'strong' form by Weismann,
suggests that the origin of adaptations lies in natural selection
acting on hereditary variations which are in their origin non-
adaptive. All other theories prove either to be versions of one of
these two, or, as in the vitalist theories of Bergson and Teilhard
de Chardin, to be untestable and therefore to be judged as
myths rather than as scientific theories. The importance of
molecular genetics for evolution theory is that it enables us to re-
formulate the Lamarck *vs.* Weismann argument in chemical
terms.

I have therefore rewritten the chapter on Heredity, and
added chapters on the molecular aspect of the Weismann–
Lamarck argument and on molecular evolution. I have taken
the opportunity to add sections on two topics – the evolution of
altruism, and the effects of artificial selection for changing
patterns – which have interested me recently, and to make a
number of other minor alterations.

Preface to the Third Edition

This is an exciting time for students of evolution. Advances in molecular biology have made it possible for the first time to measure the rates at which genes evolve, and have provided us with more powerful tools for studying the genetic variability of wild populations. The use of the microscope in the search for fossils in pre-Cambrian rocks has extended the period for which we have direct evidence of evolution by a factor of five; at the same time advances in genetics and cell biology have given us a new insight into the events which may have been taking place in that early period. The problem of the origin of life is ceasing to be a matter for speculation alone, and is becoming a subject for experimental investigation. At the other end of the time scale, knowledge of our own recent ancestry is growing, and comparative studies of the social life of our primate relatives is stimulating discussion about how our ancestors may have lived.

When I wrote the first edition of this book, fifteen years ago, I had an ideal reader in mind; it was myself aged twenty, when I had a great curiosity about evolution but no formal training in biology. My aim is still the same. I want to cover all important aspects of the theory of evolution, and to do so without oversimplifying and without avoiding difficulties; at the same time I would like to be comprehensible to anyone who is prepared to make the effort to follow me.

The book has been drastically revised since the last edition in 1966. New chapters have been added on the origin and early evolution of life, on the structure of chromosomes and the control of gene action, and on protein polymorphism. The chapters on molecular evolution and on the evolution of man have been completely rewritten, and smaller changes have been made to the remaining chapters.

To help those who may wish to chase up references to particular pieces of work, I have given the author's name in the text, although in the great majority of cases the original reference appears in one of the books listed under 'Further Reading'.

I am again indebted to many colleagues. Dr Lynn Margulis has read Chapter 6, Dr Richard Andrew Chapter 19, and Dr Bridgid Hogan Chapters 5, 6 and 7. All of them have made helpful suggestions, and have prevented me from making some errors; it is, however, only fair to them to add that in some cases I have not taken their advice.

Introduction to the Canto Edition

I was at first reluctant to agree to the re-issuing of this book. It was first published in 1958, and last brought up to date in 1975. A lot has happened since then. Two things have changed my mind. First, there is no other account of evolutionary biology available which is at the same time written for a non-professional readership, and which covers the whole field, from the origin of life to human evolution, and from molecular biology to animal behaviour. Second, I find on re-reading it that the picture it presents is close to the one I would paint if I were to start afresh, and write a wholly new book.

All the same, much has been discovered in the last twenty years. I now attempt to summarize some of these additions.

(i) *Molecular Biology*

Rapid advances in this field have transformed many branches of biology, and evolution theory is no exception. The account of molecular biology in Chapter 5 is still adequate, but there is more to say about the application of these facts.

(*a*) *Molecular Weismannism.* The central idea that underlies this book is that the origin of new heritable variation is not adaptive. Most new mutations are harmful. If evolution leads to adaptation, as obviously it does, it is because selection establishes the small fraction of mutations that are adaptive. The alternative, 'Lamarckian', view is that individual organisms adapt during their lifetimes, and pass those adaptations on to their offspring: the so-called 'inheritance of acquired characters'. In

1

Chapter 4, I gave a brief explanation of how the Weismannist view had been given a molecular interpretation in the 'central dogma' of molecular biology: acquired characters are not inherited because information cannot pass from protein to DNA, but only from DNA to protein.

Since 1975, two groups of facts have emerged that might seem to challenge the central dogma. The first concerns 'reverse transcription'. As explained in chapter 5, information passes first from DNA to an intermediate, messenger RNA, and then to protein. The first of these stages, from DNA to RNA, is called transcription. It depends on the pairing of complementary bases, just as does the replication of DNA (p. 71). It turns out that the transcription step is reversible: information sometimes passes from RNA to DNA. This process of 'reverse transcription' is of great practical importance – for example, it is essential for the replication of the virus that causes AIDS. But, despite claims to the contrary, it has no relevance to the central dogma. When I wrote (p. 80) that it is difficult to see how the flow of information could run backwards, the step I had in mind was that from RNA to protein. There is still no reason to think that this step can be reversed. Of course, even if the protein-to-RNA step could be reversed, the inheritance of an acquired character would also require that the change in phenotype be translated into a change in a protein, and in most cases it is hard to see how this could happen.

A second group of facts is much more controversial, and could lead to a bigger revision of our views of evolution. Cairns, and more recently Hall, have studied mutations in bacteria that are starved, and therefore not growing. For example, bacteria need the amino acid, tryptophane, in order to grow. Most can make it for themselves, but some 'trp⁻' bacteria have undergone a mutation in a gene coding for an enzyme that helps to make tryptophane: these can grow only if they are supplied with tryptophane. Cairns and Hall measured the rate of 'back mutation' of trp⁻ bacteria to a state in which they can again grow in the absence of tryptophane. They found that the rate is higher in cells that are starved of tryptophane, and cannot grow, than it is in growing cells.

By itself, this finding is interesting, but does not challenge the idea that mutation is non-adaptive. It would be explained if a cell which is in difficulties, and cannot grow, increases the mutation rate of all its genes. This would be a sensible thing to do: if in trouble, try anything. However, Cairns and Hall go further, and claim that the mutation rate increases only, or at least mainly, in those genes which, if they mutate, will help the cell to resume growth – in this case, the gene that synthesizes tryptophane. This claim is still highly controversial. If it does turn out to be true, how could it be explained? The difficulty if this: how does the cell 'know' which genes to mutate? Several mechanisms have been suggested. The one that seems most likely to me is as follows. Not all genes are transcribed (that is, copied into RNA) all the time. Suppose that genes that are being copied are more likely to mutate than those that are not. A cell that needs tryptophane to grow will be desperately trying to synthesize it: therefore, the relevant gene will be switched on, even if it is no good (the control mechanism involved is explained on p. 123). This could explain apparently adaptive changes in the mutation rate. I must emphasize that there is as yet no evidence that this is the correct explanation. But the idea is testable. I offer it to make the following general point. If we are faced with an apparent case of adaptive mutation, we now know enough molecular biology to seek a mechanism to explain it.

I have spent some time on this example because the 'Weismann *vs.* Lamarck' argument remains crucial for evolution theory. The snag with Lamarckian explanations is that there seems to be no way in which an organism could recognize the adaptive changes – and only the adaptive ones – it had undergone, and convert them into corresponding changes in DNA. It is too early to be sure of the significance of these recent bacterial experiments. It may be no more than that cells in difficulties increase the rate of mutation in a non-specific way. If, as seems possible, something more is happening, it will be fascinating to find out how it works. In any case, the process can only help a cell to meet an immediate molecular problem: it could not lead to morphological or behavioural adaptation.

(*b*) *Sequence Data and the Mechanism of Evolution.* One major technical advance has been in methods of determining the sequence of nucleotides in DNA. This information has been useful to evolutionary biologists in two main ways: in determining relationships, and in analysing mechanisms of change. To shed light on evolutionary mechanisms, we need the sequence of the same gene from a number of closely related individuals – members of the same species, or of similar species. Such information is only just beginning to be available because, understandably, molecular biologists have preferred to sequence a gene as different as possible from anything that has been sequenced before.

The value of having a number of sequences, or other molecular information, from related individuals is that it can tell us about the nature of the variation upon which selection can act, the kind of changes that occur, and the extent to which genes are exchanged between populations. Some examples will make these points clearer. In Chapter 12, I discussed the idea that the evolution of social behaviour depends on genetic relatedness. Molecular methods have been used to measure relatedness in animal societies. In some cases, it has been shown that the degree of altruism displayed towards others varies in the predicted way with relatedness. In the comparable problem of parental care, one would expect the amount of paternal care to vary with confidence of paternity: no increase in fitness follows from caring for unrelated offspring. Molecular studies of birds that form monogamous pairs have shown that the frequency of 'extra-pair copulations' is surprisingly high. In some cases, males do reduce their care of the young if their mate has had opportunities to copulate with another male.

The evolution and maintenance of sex has received increasing attention. There are two contexts in which molecular information is crucial. One concerns the longevity of clones (that is, asexually reproducing lineages). It is accepted that the ancestors of animals, plants, and fungi were sexual, but in all three groups some lineages have wholly abandoned sex. For how long can a lineage survive without sex? As yet, we do not know whether any animal clones are really old – millions rather than thousands of years. The obvious candidates are the Bdelloid rotifers (small multi-cellular fresh-water 'wheel animalcules'), a whole sub-order in which no one has ever seen a male. Are they a genuinely ancient clone, many millions of years old, or have they invented some alternative to males as a means of exchanging genes? We should soon know.

A second question concerns the prokaryotes (bacteria and blue-green algae). These do not have the classical sexual processes of meiosis followed by gamete fusion, but, at least in the laboratory, there are ways in which single genes, or parts of genes, can be transferred from one cell to another. Have these parasexual processes been important in the evolution of bacteria? Sequence analysis has shown that gene transfer has been crucial in the evolution of drug resistance, and in antigenic changes that enable bacteria to escape the immune responses to their hosts. Infectious disease would be a good deal easier to cope with if our parasites did not have means of exchanging genes.

The availability of DNA sequences has had an important influence on the debate (pp. 102–6) about the 'neutral mutation theory': that is, the idea that most changes at the molecular level happen, not because they are selected, but because they are selectively neutral. If the theory is true, there are two predictions. First (p. 104), the rate of evolution of a particular gene, or region of DNA, should be constant. Second, the rate should be high for those DNA regions on which there are few selective constraints (that is, which can change with little effect on fitness), and low for highly constrained regions (that is, regions in which most changes would have deleterious consequences). If we compare the DNA sequences of the same gene in related individuals, we can distinguish two kinds of change, 'synonymous' and 'substitutional': a synonymous change is one which, because of the redundancy of the genetic code (p. 91), causes no change in the amino acid, and a substitutional change is one that does cause the substitution of one amino acid for another. We would expect there to be greater selective constraints on substitutional changes (although it has turned out that even synonymous changes can be selected for or against, because some codons are translated more slowly than others), and hence, if the neutral theory is correct, the rate of synonymous change in evolution should be higher. This is in fact the case. However, sequence analysis has provided evidence that, in at least some genes, most amino acid changes in evolution are selective rather than neutral. Perhaps the strongest evidence is that, in the ADH gene of *Drosophila*, there are more amino acid differences between related species than would be predicted on the neutral theory, knowing that there is little variation within species.

Molecular data have been used extensively in determining population structure. For example, some 300 killer whales have been studied behaviourally for twenty years off Vancouver Island in British Columbia. They fall into two groups, one of which follows the seasonal salmon migrations, and the other of which feeds on marine mammals. The groups differ in DNA sequence to an extent as great as that which separates killer whales from the Pacific and Atlantic, suggesting that, although they inhabit the same region, they do not interbreed. Information of this kind is of obvious value in conservation. It is also relevant to the origin of new species among mammals: although the two killer whale populations should probably not be regarded as different species, the difference in behaviour could be a first step in the speciation processes.

(c) Molecular Data and Phylogeny. A curious omission from earlier editions of this book is the lack of any discussion of the theory of classification. I spent some time on the nature and origin of species, but said little about classification at higher levels, beyond saying that a hierarchical classification (species – genus – family – order – class – phylum) fitted the observed pattern of variation, as would be expected on evolutionary grounds. As to how classification should be carried out, I said only that species that resemble one another in many characteristics should be grouped together. I had not at that time digested the ideas of Willi Hennig, whose book on systematics, published in German in 1950 and translated into English in 1965, has become the orthodoxy among taxonomists. The application of his ideas owes a lot to molecular data, and to computers, but it will be clear from the date of publication that neither contributed to their origin. I will explain them with a morphological example. The fact that horses and zebras both have a single toe is regarded as evidence of close relationship, whereas the fact that humans and lizards both have five toes is not. Why should this

be so? The reason is that, for land vertebrates, to have five toes is the primitive condition, and to have a single toe is a derived character. Resemblance in a derived character is good evidence of relationship, but resemblance in a primitive character is not. The principle is a good one, but how does one decide which are primitive and which derived characters? Sometimes one can get an idea of the primitive state from the fossil record, or from development (p. 311), but the most widely applicable method is the use of an 'outgroup'. For example, when classifying the Perissodactyls (horses, tapirs, rhinos, etc.), one would take as an outgroup some other mammal. The perceptive reader will notice that there is an element of circularity here: how does one choose a suitable outgroup until one knows the classification?

The relevance of molecular data is that they provide a vast number of additional characters that can be used in classification. Given computers, these data can be pressed into service. Do molecular data have any intrinsic advantage, other than sheer volume? Two features perhaps make them peculiarly useful. The first is the non-adaptive nature of many molecular changes. Adaptive characters may evolve independently in different lineages. Thus a single toe is an adaptation for running fast in open country: it evolved not only in horses but also in an extinct group of South American mammals, the Litopterns. A second feature of molecular changes, causally connected to their frequently non-adaptive nature, is their approximately constant rate. A molecular classification, therefore, may give, not only a reliable phylogeny, but also an approximate dating of the times of divergence of the various lineages.

What has emerged from molecular phylogenetic studies? In general, they have confirmed classifications made on morphological ground. Many details have changed, and doubtful points have been cleared up, but the basic picture remains unchanged. Perhaps the most important contribution has been to the relationships between major groups – phyla and kingdoms – which are so different that morphological information is unhelpful. The concordance between molecular and morphological phylogenies is to be expected if the theory of evolution is true, and inexplicable otherwise. An important novelty concerns the evolution of proteins themselves. It was already familiar from morphological studies that new organs, with new functions, do not emerge from nothing, but by modification of already existing organs with different functions. Arms and legs are modified fins, wings are modified arms, feathers are modified scales, jaws are modified gill arches, and the swim bladders of fish are modified lungs (although, as it happens, Darwin thought it was the other way round). The same picture holds for proteins, as is shown by the similarity of sequence between proteins with quite different functions. For example, lysozyme, a bacteriocidal protein present in tears, has sequence similarity to an enzyme that helps to make lactose in the mammary glands. This could not have been predicted, but has been explained retrospectively by saying that the first protein breaks a chemical bond similar to that made by the second.

(*d*) *Selfish and Ignorant DNA.* One surprise has been the discovery that a large proportion of the DNA in eukaryotes is never translated into protein. In humans, as little as ten per cent of the DNA is translated, and the proportion is still lower in newts, lungfish and lilies. Some of the untranslated DNA performs a useful function: it may regulate gene action, or be transcribed into RNA that plays a role in protein synthesis or in other ways. But the vast majority probably does nothing useful for the organism at all.

To understand why this is so, remember that the nucleus of a cell is packed with enzymes that replicate DNA, and others that cut it and splice it together again, the function of the latter being to repair damaged DNA, and to recombine chromosomes (p. 61). Hence a DNA molecule in the nucleus, particularly if it is inserted into a chromosome, will be replicated, even if it performs no useful function. It helps to think of the additional DNA as falling into two categories, 'ignorant' and 'selfish'. The ignorant DNA does not have any special sequence that ensures its survival. Often it consists of short sequences of five to ten nucleotides, repeated over and over again. It is just there, and replicated because it is there. In contrast, selfish DNA has an evolved sequence that ensures its own increase. For example, in the chromosomes of wild *Drosophila melanogaster*, there are some fifty 'P factors'. These are regions of DNA some 3000 nucleotides in length which are transcribed and translated into two proteins. One of these causes additional copies of the P factor to be inserted elsewhere in the chromosome set – a process called transposition – and the other controls the process. In most populations, P factors cause no particular harm, although it must cost the fly something to replicate all this useless DNA. But if strains of *Drosophila* with and without P factors are crossed, the control of transposition breaks down, causing death and infertility.

Such 'transposable elements' are universal. To give a second example, there are some 400,000 copies of the Alu element, 282 nucleotides long, distributed throughout the human genome, amounting to about five per cent of the DNA in the nucleus. The existence of transposable elements raises a problem for evolutionary biology. As the P factor example shows, an element that transposes too successfully can damage the organism. We are therefore faced with another example of selection operating on two levels. On pp. 193–200, I discussed the problem of group selection and the evolution of social behaviour: why do individuals cooperate in animal societies, despite selection for selfish behaviour? We are now faced with an analogous question: why do the genes in an organism cooperate to ensure the survival of the organism, despite selection for selfish replication? The question is easier to ask than to answer.

One last comment on molecular biology: the prospects discussed on the last page of the book come ever closer.

(ii) *Replicating molecules*

It is now possible to study evolution in a test tube, in the absence of any living organisms. A test tube is prepared containing the four nucleotides from which RNA is synthesized, a 'primer' molecule of RNA, and an enzyme, Qβ replicase, which copies RNA molecules. The enzyme repeatedly copies the primer, using the nucleotides provided. After some hours, when many copies exist, a drop of the solution is transferred to a second tube, also containing enzyme and nucleotides, but not, of course, a primer, because RNA molecules ready to be copied are already present. The process can be repeated as often as one wishes. If replication was precise, this would merely produce many copies of the original primer. But replication is not perfect. Every time a new nucleotide is added, there is a chance of about 1/1000 that it will be 'wrong': that is, it will not be complementary to the nucleotide in the strand being copied. Other errors, or mutations, lead to changes in the length of the RNA molecule. Since some RNA sequences are replicated more rapidly than others, there is a process of evolution by natural selection. For a given set of physical and chemical conditions, the end point of this evolutionary change is repeatable – usually an RNA molecule some 200 nucleotides long. There is, apparently, some unique 'best' sequence, and natural selection can rather rapidly produce a population consisting of molecules with this optimal sequence, or one very like it, regardless of the sequence of the original primer.

Of course, these experiments are not an answer to the question of how life originated. Conditions in the test tube differ from those in the primitive ocean in one crucial respect: there could not have been any $Q\beta$ replicase molecules present in the primitive soup. Nevertheless, the experiments are interesting for two reasons. First, they demonstrate how, once replication has arisen, natural selection can generate structures which, without it, would be wildly improbable. Thus there are 4^{200}, or 10^{120}, different RNA molecules 200 nucleotides long, yet natural selection can repeatedly produce one specific sequence in a few days. The experiments are also important for a practical reason. Modifications of this procedure may make it possible to use natural selection to produce enzymes with specific desired activities.

(iii) *The Origin of Life*

In existing organisms, nucleic acids, DNA or RNA, act as carriers of genetic information, and proteins act as enzymes responsible for metabolism. This led to a 'chicken and egg' problem: did nucleic acids or proteins come first? How could nucleic acids be first, if there were no enzymes to replicate them? How could proteins be first, since they cannot replicate, and so will not evolve by natural selection? However, we now know that some RNA molecules have enzymic activity. The chicken and egg problem therefore disappears. We now imagine an 'RNA world' in which the same molecules acted as enzymes and as templates carrying genetic information.

Serious difficulties remain. Was RNA the first replicating molecule, or could there be something simpler, but still depending on complementary base pairing for replication? How did the genetic code originate?

(iv) *Behavioural Ecology*

The topics discussed in chapter 12 have become a major preoccupation of evolutionary biologists, giving rise to the discipline of behavioural ecology. I start by discussing two conceptual topics. The first is the introduction of evolutionary game theory. Game theory was first applied to economics, to discuss what rational people should do when playing a 'game', defined as an interaction in which different participants want different outcomes. Evolutionary game theory asks an analogous question. How will a population evolve if the best thing to do (or, if you prefer a more formal statement, the fittest phenotype) depends on what others do (that is, the phenotypes of other members of the population)? In other words, it is a way of thinking about frequency-dependent selection (pp. 181–3). The central idea is an 'evolutionary stable strategy', or ESS. An ESS is a strategy, or phenotype, with the following property: if almost all the members of a population adopt that strategy, no alternative strategy, arising by mutation, can invade the population. In other words, no other strategy can have as high a fitness: an ESS is a strategy that does well when surrounded by copies of itself. Clearly, if a population comes to consist of individuals adopting the ESS, it will cease to evolve. Evolutionary game theory, therefore, does not help much in understanding change, but, since most populations have had time to come close to the optimum for the environment in which they live, it does help us to understand the selective forces and constraints that have shaped the animals and plants we see around us.

The idea will be illustrated by the evolution of the sex ratio – that is, the relative numbers of males and females. In its simple form, this problem was solved by R. A. Fisher, long before the introduction of evolutionary game theory, but it is an admirable example of an ESS. Suppose that females could choose the sex of each child (an exactly similar argument applies if males could choose). Which sex should a female choose? The Darwinian argument says that she should choose whichever sex maximizes the number of her grandchildren. But this depends on the sex ratio in the population, and hence on the choices made by other females: this is why it is a game. Thus if there are more females than males, males will on average have more children, and vice-versa: this follows from the fact that every child has one father and one mother. Each female, therefore, should choose to have children of whichever sex is the rarer. It is easy to see that the only stable state is one with equal numbers of males and females. Either each female should choose the sex of each child by tossing a coin – in effect, this is what happens in most species – or half the females should produce only sons and half only daughters. A 1:1 sex ratio is the ESS.

Of course, in most species females do not 'choose' the sex of each child: it is determined by the segregation of X and Y chromosomes in meiosis (p. 62), a process that does in fact produce equal numbers of sons and daughters. But if the evolutionarily stable sex ratio was *not* 1:1, I have no doubt that some other mechanism of sex determination would have evolved. It is important that the term 'strategy' does not imply conscious choice: an ESS is a phenotype that will evolve under natural selection. In some animals (for example, the hymenoptera – p. 196), sex is not determined by an X–Y mechanism, but by whether or not the egg is fertilized. In a sense, a female can choose the sex of each offspring. It turns out that, when the conditions assumed in Fisher's argument do not hold (notably, that males have free access to females, and vice-versa), the sex ratio departs from 1:1 in just the way predicted by game theory. For example, in parasitic wasps which lay their eggs in a caterpillar, and in which the emerging males mate among themselves before dispersing, there is a great excess of females: a female need only produce a few sons to ensure that all her daughters are mated.

This way of thinking has been widely applied to the evolution of animal behaviour. Topics that have been analysed in this way include territorial behaviour, sexual selection, fighting behaviour, the conflict between parents and offspring, foraging, male mating strategies, and many others. The method is not confined to behaviour but is relevant whenever fitnesses are frequency-dependent. Game theory has been applied to plant growth, and to the evolution of viruses. These last two examples reinforce the point that the theory makes no assumption of rational choice.

The second conceptual topic is the evolution of animal signalling. It was clear from the work of the ethologists (pp. 209–10) that animals do make ritualized signals, which elicit specific responses. There is, however, a theoretical difficulty in understanding the evolution of such signals: why are they believed? Consider, for example, the evolution of a specific threat display (p. 210). There is no point in making such a display unless it causes one's opponent to back down. But if giving the display has that effect, why not give it even when you do not intend to attack? But if everyone gives the signal, regardless of their intentions, no one will believe it, and the whole signalling system breaks down.

One way out of this difficulty is to argue that signals are really 'assessment signals'. For example, the signals (roaring, and 'parallel walking') made by red deer stags to one another during the rut carry accurate information about size and physical condition: therefore they cannot be faked. However, this may not be the whole story. Zahavi has suggested that reliable signals are costly to make, and it is this cost that ensures that they are honest. I was one of those who was slow to accept this argument, partly because it was expressed in a verbal rather than a mathematical form – I find verbal arguments hard to follow. There is, therefore, a certain irony in the fact that the correctness of Zahavi's argument has been confirmed by Enquist and Grafen, using mathematical methods (game theory) that I was partly responsible for introducing. The matter is still controversial: there is need for empirical tests designed to test the various theories. It is also of some importance, not least in thinking about human evolution. At some point, our ancestors evolved a communication system, language, whose reliability does not depend on the signal being costly.

The study of sexual selection, and in particular of female choice, has attracted increasing attention. This marks a break with the past. When Darwin published his ideas about sexual selection in 1871, the notion of female choice was not well received. I think that this was because the idea of choice did not appeal to those who were thinking about animal behaviour at that time. Their aim was to reduce behaviour to a series of 'tropisms', similar to the responses of plants to light and gravity. Animals were thought of as simple machines, with nothing in their heads: a similar outlook, behaviourism, dominated human psychology for many years. Such an outlook left little room for choice. In animal behaviour, it was replaced in the 1940s by the ethological outlook, pioneered by Lorenz and Tinbergen: animals were born with the ability to respond in complex ways to complex stimuli. Surprisingly, however, this did not at once lead to renewed interest in Darwin's idea of female choice. When, in 1956, I published some observations on female choice in *Drosophila* (pp. 212–14), few people took any notice, and those who did rejected my interpretation.

At that time, it was accepted that females – and males – made choices during courtship, but choice was thought only to ensure that a female mated with a male of the right species. This is not surprising, since much of the work of evolutionary biologists in the period of twenty years or so before the first edition of this book was published was concerned with the nature and origin of species, as is reflected in the four chapters (13–16) devoted to these topics. In fact, one function of courtship is indeed to ensure that inter-specific matings are avoided. But it is now clear that females do choose between conspecific males.

The swallow is one species in which female choice has been extensively studied in the field. Swallows are migratory, and monogamous for a breeding season. The males return first to the breeding grounds. Females pair soon after they return. Möller has shown that the first females to return mate preferentially with the males with the longest tails. These pairs produce, on average, more young that fledge: it is not clear whether this is a direct result of starting to breed early, or whether it depends on the greater fitness of the females (because, perhaps, fitter females return earlier), or of the males (males with longer tails are, perhaps, fitter). The preference of females for males with longer tails is confirmed by the fact that females paired with shorter-tailed males are more likely also to copulate with a male to which they are not paired, and, if they do, to choose a male with a longer tail. So there is evidence of female choice, and evidence that males with longer tails are benefiting. But are the females getting anything out of it?

One piece of evidence that they may be doing so comes from experiments in which tail length has been artificially altered. If the terminal parts of the tail feathers of a male are removed, and then glued back on again without altering the length, the bird is as likely to return next year as if no operation was performed, showing that the operation itself has no harmful effect. Naturally long-tailed males are as likely to return as naturally short-tailed males, but if a short-tailed male is converted into one with a long tail, its chances of returning are reduced. This suggests that too long a tail may be a handicap, but one which naturally long-tailed males are able to support. If so, these males are indeed fitter, and females which choose them may be benefiting, either in better care, or better genes, for their offspring. If so, this is an example of Zahavi's idea of honest signals being costly. These experiments show how difficult it is to sort out what is happening, but also that real progress is being made.

There has, during the last twenty years, been a big change in the rigour with which ideas about the evolution of behaviour are tested. It is more difficult than it is in genetics, for example, to design experiments to decide between alternative hypotheses, and more difficult to repeat experiments which may have taken years of field work. Nevertheless, standards have improved, partly because, with more people working on behaviour, there is a greater chance that conclusions will be challenged, and observations repeated. There has also been progress in the 'comparative method', which is my next topic.

(v) *The Comparative Method*

Suppose we want to know what have been the selective forces responsible for the evolution of some trait. One way of finding out is to look at the distribution of that trait among species. For example, in many primates (monkeys and apes), males are bigger than females – the gorilla is an extreme example – whereas in others, such as gibbons and marmosets, there is little difference. Two selective explanations have been suggested. One is that males are larger because they must fight for access to females. The other is that it pays the members of a pair to be different in size because then they can exploit different food resources within a territory. If the former explanation is correct, we would expect the size difference to be greater in those species in which, in the breeding group, there are several adult females to each male. In contrast, if the second explanation is correct, we expect the difference to be greater in species that form monogamous pairs. In fact, in monogamous species (gibbons and marmosets are examples) there is little size difference, whereas there is usually a substantial one in species with several adult females to each male. It follows that competition for females, rather than sharing of food resources, is a more likely explanation for size dimorphism in primates.

Several points need to be made. The first, and most obvious, is that one must not choose one's species to prove one's case. It may not be possible to include data for all species, because the facts may not be known, but at least one must include all species for which data are available. Second, a conclusion that holds for primates may not be true for other taxonomic groups. For example, in predatory birds females are larger than males. This is not because females compete for males: the true explanation is still obscure.

A less obvious but equally important point is that the observed correlation should not be an accident of history. Thus suppose that, in the primate example, it turned out that all monkeys were monogamous and showed little size dimorphism, whereas all apes were dimorphic and polygynous. Then the association between polygyny and size dimorphism might merely be a relict of something that was present in the ancestor of the apes: it would not prove a causal connection between them. In this particular example, the association cannot be explained in this way (for example, gibbons are apes, and marmosets are monkeys). But in many cases it is not obvious whether an association is causal or an accident of history. To decide, two things are needed. First, one must know the relationships of the species in the sample: this is one reason why the construction of reliable phylogenies is important. Second, the data must be analysed using appropriate statistical methods. During the past twenty years, the comparative method has ceased to be a matter of looking for a few cases that fit one's favoured theory, and has become a respectable branch of science.

(vi) *The Fossil Record*

One of the few actual errors that I am aware of in earlier editions concerns the evolution of *Gryphea*. It is now thought that *Gryphea* did not evolve independently from oysters on many occasions, but happened once only, and that the repeated replacement of oysters by *Gryphea* as sediments became more muddy in particular locations represents, not evolution *in situ*, but the replacement of oyster species by immigrating *Gryphea* species. This does not affect the argument I was making against 'racial suicide' – indeed it rather strengthens it – but the error needs pointing out.

One change in the practice of palaeontology has been the increasing use of quantitative methods. In earlier times, the main aim was to fill in gaps in the fossil record. That continues to be important, but, as the record becomes more complete, attention has turned to other questions. How many species were there at different times, and in what habitats? What were the rates of species origin and extinction? Does the likelihood of a species going extinct depend on particular features of its biology? How fast do individual lineages evolve, and is the rate uniform? For those interested in the mechanisms of evolution, these studies are more likely to be informative than the search for missing links. This has tended to bring palaeontologists back into the mainstream of biology.

A claim of which perhaps too much has been made concerns 'punctuated equilibria'. Gould and Eldredge suggested that evolution has not proceeded at a uniform rate, but that most species, most of the time, change very little, and that this condition of 'stasis' is occasionally interrupted by a rapid burst of evolution. I do not doubt that this picture is sometimes, perhaps often, true. My difficulty is that I cannot see that it makes a profound difference to our view of evolution. You can judge this for yourself by reading my discussion of rates of evolution (pp. 276–85). Certainly, there is nothing in the data to suggest that any special processes are involved during these periods of rapid change: it is well to remember that the rates are probably still small compared to those that can be produced by artificial selection.

Perhaps the most dramatic new claim is that the mass extinction of species at the end of the Cretaceous was caused by a collision of a meteorite with the earth. Although (p. 296) I mentioned the extinction of the archosaurs, I failed to make it clear that, at the same time, there was also a mass extinction of marine species. Although there is still controversy, I find the evidence for a meteorite collision persuasive. There have been other mass extinctions, but it is not clear to me that they were similarly caused.

An important addition to our knowledge of the kinds of animals and plants that existed in the past concerns the first great explosion of multicellular animals in the Cambrian (p. 120). Most fossils from that time tell us only about hard parts

– shells, spines and carapaces. But in the Burgess shale, information about soft parts is beautifully preserved. These fossils have been known for over fifty years, but recently they have been re-examined. It is now clear that there existed in the Cambrian a very wide array of forms, some of which may differ in their basic body plan from anything alive today. It also seems likely that, with a few minor exceptions, all the body plans that exist today were already present in the Cambrian. From the point of view of body plans, the full range of variation arose early in the history of multicellular life.

Some biologists whose judgement I respect doubt this conclusion. The snag is that we interpret these very early animals in the light of what we know of existing ones. For any particular Cambrian animal, we seek similarities to an existing one. If we find such similarities, we place the fossil in the same phylum as the existing animal: if we fail to see any similarities, we erect a new phylum to hold the new fossil. In this way, we identify many phyla in the Cambrian, not all of which exist today. Perhaps if we looked at the Cambrian fauna with an open mind, and not in the light of what has happened since, we might conclude that they are really not all that variable. There are also doubts about the reality of the 'body plans', whose identification was the major goal of classical comparative anatomy. Despite these difficulties, I am inclined to accept that these early faunas contained animals with a very wide range of body plans. Of course, this may merely reflect the fact that I was trained in zoology at a time when comparative anatomy was still the dominant discipline.

It is important to understand what is being said here. In terms of variation in ways of living, organisms today are enormously more variable than they were in the Cambrian. There were then no land animals or plants, no animals that could fly, and, so far as we know, no social animals, and none that could echolocate, or make webs, or talk. But these extensions in ways of life have been achieved without change of body plan. Birds, bats and humans still have the basic vertebrate body plan of a backbone, dorsal hollow nerve cord, two pairs of limbs, and so on, which originally evolved in our swimming ancestors. What we have to explain, therefore, is why such a wide range of body plans evolved in or before the Cambrian, and few if any new plans have evolved since then.

In fact, there seem to be three things we have to explain:

(i)   Some time before 600 million years ago, there evolved the capacity to develop a complex body plan, with many different kinds of differentiated cells, arranged in a complex and repeatable structure. Once this capacity existed, a number of body plans evolved.

(ii) Once evolved, the basic body plan did not change. New ways of life evolved by changing the function of pre-existing parts.

(iii) Few if any new body plans have arisen since this early radiation.

I will take these points in reverse order. The absence of any new body plans need not indicate any loss of evolutionary potential. It is more plausible that, once a great variety of complex animals existed, further attempts to evolve a complex body plan from scratch were inhibited by competition. In the same way, we think that the origin of life itself was a unique event, because once living organisms existed they would rapidly destroy any later beginnings. Turning to point (ii), the failure of body plans to change arises from an argument on p. 312: structures that have lost their original adult function persist because they play a causal role in development.

The hard question concerns point (i) above. What invention or inventions made possible the development of complex multicellular bodies? Essentially, this is a problem for developmental biologists.

(vii) *Development and Evolution*

I have long thought that an understanding of development should illuminate evolution. After all, our basic theory is that changes in genes cause changes in adult phenotypes, and that selection acting on those phenotypes determine which genes persist, and which are eliminated. But the problem of how genes specify adult phenotypes remains a black box, whose working we do not understand. Progress has been slow, but recent work, particularly on *Drosophila*, on the nematode worm *Caenorhabditis*, and on the plant *Arabidopsis*, are, I hope, bringing about a breakthrough, although it is one whose evolutionary implications are still hard to see. The method has been to identify the genes concerned with particular stages in development (for example, segment formation in *Drosophila*, and flower morphology in *Arabidopsis*), determining their nucleotide sequence, identifying the exact time and place when they are active, and studying their interactions.

To summarize this work would take a whole book – and one that I am not competent to write. I have room only to mention one topic, gene regulation, and one recent finding. I discussed gene regulation on pp. 122–4. I explained how, in bacteria, genes can be switched on and off. But, in this system, to switch a gene on, and to keep it switched on, requires the continued presence of an 'inducing' molecule. Therefore the mechanism illustrated in Figure 14 is not sufficient to explain the more long-lasting changes that occur in differentiated cells (p. 124), whereby cells of particular types 'breed true': epithelial cells, when they divide, give rise to epithelial cells, fibroblasts to fibroblasts, and so on. This does *not* require a change in nucleotide sequence in DNA, but some kind of mark or imprint on the DNA which is copied when the cell divides. One such mark, known to be used both in bacterial and eukaryotic cells, is the methylation of particular bases. Thus there is no difficulty, in principle, in understanding how the state of activity of a gene can be altered, and how that changed state can be transmitted in cell division. It is harder to see what had to be invented by the first multicellular animals, since a possible mechanism already exists in the bacteria. Of course, regulation must be much more complex in animals: the activity of a gene may depend on the position of the cell in the body, on the kind of cell it is, on the stage of the cell cycle, and on the age and sex of the animal.

The one new finding I want to discuss concerns the 'homeotic' genes of *Drosophila*, discussed on p. 318. Mutations in these genes cause the development of the 'wrong' segmental appendages: a leg where there should be an antenna, or a wing where there should be a haltere. A number of such genes have now been isolated, and their nucleotide sequences and times of action determined. All of them contain a 'homeobox' region of sixty amino acids (180 nucleotides), which varies between genes, but is sufficiently similar to indicate common ancestry. Most surprisingly, a similar set of genes has been found in the mouse. Still more surprising, for each *Drosophila* gene, there is an homologous mouse gene with a high degree of sequence similarity. If one then identifies the place of action of these genes along the antero-posterior axis of the body, it turns out that the most anterior-acting *Drosophila* gene is homologous to the most anterior-acting mouse gene, and so on along the axis of the body.

What do these facts tell us about evolution? At first sight, they suggest that the common ancestor of mouse and *Drosophila* was a segmented animal, but there are good morphological grounds for thinking that this is not the case. At most, the common ancestor was a bilaterally symmetrical animal, with a head, a middle, and a back end. Presumably, the series of homeobox genes was already present in that ancestor. It turns out that the homeobox region (but not the series of homologous genes) goes back further than that. A gene with a homeobox region determines mating type in yeast. That is, it determines the difference between two type of cells, which will fuse with one another but not with themselves. To do this, the gene must control the activity of a number of other genes, which specify the cell surface properties that determine fusion, the production of pheromones (chemical attractants), and the reception of those attractants. This may be a clue to what the homeobox region is for. Genes containing it control the activity of a number of other genes, whether those genes act in the development of cells of a particular mating type, or of legs rather than antennae.

Returning to the question of what had to be invented before complex multicellular animals could evolve, the answer must still be that we do not know. The homeobox story, however, shows that one regulatory system was already present in the common ancestor of arthropods and vertebrates, that it was initially concerned with the differentiation of head, middle and tail, but was later used to control other differences. We can hope that, during the next twenty years, as our knowledge of developmental genetics grows, we shall at the same time gain more insight into the great Cambrian explosion of body forms.

(viii) *Human Evolution*

New human fossils continue to be found, and add detail to the picture give in chapter 19. For example, it is now clear (p. 335) that two species of Australopithecines, robust and gracile, did coexist in Africa. The status of *Ramapithecus* (p. 333) as a human ancestor is no longer generally accepted: the molecular evidence suggests that the divergence between apes and humans may have occurred as recently as five million years ago. The basic conclusion that bipedal locomotion preceded any great increase in brain size still holds. The study of tools, in Africa and Europe, shows that a dramatic advance took place, in sophistication and diversity, some 200,000 years ago. It is tempting to suggest that this was triggered by the evolution of language. I will return to this possibility, but first I describe some findings from molecular biology.

In a famous paper, Cann, Stoneking and Wilson constructed a human phylogeny, using data for DNA extracted from mitochondria (see p. 117). The essential point here is that mitochondria are inherited only in the female line. Hence all existing humans are the end points of a branching tree, representing lines of female descent: it was this tree that the authors tried to construct. Clearly, such a tree must have a root, representing the female from whom all of us inherited our mitochondria. This point has caused a lot of confusion. But if each of us were able to trace back our material ancestries, these would ultimately converge on one female: of course, that convergence could be so long ago that we would not regard that female as human. Knowing the average divergence in nucleotide sequence between existing humans, one can calculate how long ago that female lived, provided one can also estimate the rate at which human mitochondrial DNA evolves. The authors estimated the latter rate by measuring the divergence that has arisen among the aboriginal inhabitants of Australia, New Guinea, and America, knowing approximately the dates at which these regions were first colonized. The answer is that our common female ancestor lived some 250,000 years ago: this estimate may be out by a factor of two, but probably not by more than that.

Before discussing the implications of this estimate, there is one point that must be made clear. It is *not* the case that, at that time, there was only one female who has contributed nuclear genes to present human populations. There were probably several thousand such females alive at that time. The claim is that only one of them contributed mitochondria. The difference is that nuclear genes are contributed by both parents, but mitochondria only by one.

So what does an estimate of 250,000 years tell us? It does help to settle one controversy. Are the existing human races inhabiting different parts of the world the modified descendants of *Homo erectus* populations inhabiting those regions up to one million years ago? Or are they the descendants of some single *H. sapiens* population that has spread across the world, eliminating the earlier *H. erectus* populations? If the estimate of 250,000 years is even roughly correct, the second alternative is nearer to the truth. One final point: in their paper, the authors argued that their data showed that this ancestral population lived in Africa. This conclusion has been challenged. I think the challenge is well-founded, although there are other data favouring an African origin for *H. sapiens*. But I do not think that the challenge affects the estimated date of our common female ancestor.

Finally, what of the origin of language? The idea that it was the origin of language that triggered the technical advance and world-wide expansion of *Homo sapiens* is attractive, although I do not see how one can be sure. I still like the idea (p. 339) that the main stimulus to the evolution of human intelligence came from social interactions. But I am unhappy with my brief remarks about language on p. 343. I have been persuaded by my colleagues in linguistics that there really is something peculiar about the human capacity to talk, and that there is a deep difference between the proto-language spoken by the chimpanzee Washoe, and by very young children, and the language of adult humans. The difference lies in grammar. Proto-language contains words that stand for objects, or actions, that can be observed, but it does not have the grammatical structure of human speech. The ability to learn to talk is not merely an aspect of general intelligence, but a peculiar, evolved ability, specific to language.

If this is correct, how did linguistic competence evolve? Unfortunately, this is not a question most linguists allow themselves to ask. There are, I think, two reasons for this reluctance. First, evolutionary speculation has a justifiably bad name among linguists. After Darwin, there were many half-baked speculations about the origin of language, and in consequence any linguist who mentions the word evolution is liable to be ostracized. Second, linguists have been engaged in a fierce argument with behaviourists, attempting to establish the uniqueness of linguistic competence, along the lines outlined above. I think they were right, but it has led them to emphasize the unique features of human language, and to deny any parallels between human and animal communication, or any possibility of intermediates between them. There are, of course, some honourable exceptions among linguists, but at present they are an embattled minority. Until evolutionary ideas again become respectable, the origins of language are likely to remain obscure. When that time does come, I have one suggestion to make. If other evolutionary novelties are any guide, linguistic competence did not arise from nothing: it is a modified version of some earlier mental capacity.

I have not tried to give references to all the statements in this introduction: professional biologists will know the work I am referring to. But you may wish to pursue these questions further. If so, I suggest the books listed in Further Reading, p. 346, none of which assume any special background knowledge.

CHAPTER 1

Adaptation

No animal or plant can live in a vacuum. A living organism is constantly exchanging substances with the environment. A tree absorbs water and salts through its roots, and loses water and absorbs carbon dioxide through the leaves. A mammal absorbs water and food substances in the intestine and oxygen in the lungs. Without these exchanges, life is impossible, although some seeds, spores, and encysted animals can maintain their organization in a vacuum, and resume their living activity when normal conditions are restored. Life therefore is an active equilibrium between the living organism and its surroundings, an equilibrium which can be maintained only if the environment suits the particular animal or plant, which is then said to be 'adapted' to that environment. If an animal is placed in an environment which differs too greatly from that to which it is adapted, the equilibrium breaks down; a fish out of water will die.

In one way or another, most biologists study this equilibrium between organism and environment. The study of evolution is concerned with how, during the long history of life on this planet, different animals and plants have become adapted to different conditions, and to different ways of life in those conditions. Some adaptations, such as the process whereby energy is obtained by fermentation, are common to almost all living things. For this very reason it is difficult to discover their evolutionary history. At the other extreme are adaptations which enable some animals or plants to live in a special way in a particular environment. In such cases we can observe differences between animals, and can often see how these

26

differences are suited to the ways in which animals live; we then have a chance to study how and why they have evolved.

The first problem to be solved, however, is this: how are we to decide that a particular feature of an animal or plant adapts it to live in a particular place? The mere fact, for example, that we find a single kind of desert plant which has succulent leaves does not prove such leaves to be adaptive. However, when we find that many different kinds of desert plants have such leaves, it is natural to look for an adaptive function, and reasonable to conclude that they can store water after rains to last during the drought to come.

This suggests the first way in which we can argue that some structure is adaptive; in effect, we say 'this is the kind of structure which in view of what we know about how things work, we would expect to perform a useful function in these conditions'. A more detailed example will make this clearer. Wild horses live in open country, and rely on their speed to escape from predators. The most striking features of a horse's legs are that there is only a single elongated toe, and that the muscles which move the foot are mainly concentrated near the hip joint, and move the foot by means of long tendons. Now in galloping a horse must accelerate and decelerate its legs with each stride, and this uses up a lot of energy. The energy expended can be reduced by lightening as far as possible the lower part of the leg, since this is the part of the leg which must be moved fastest. By concentrating the muscles in the upper part of the leg, the lower part is lightened, and the energy used up in galloping reduced.

The reason for having a single toe is less obvious. The cross-sectional area of the bones in the foot must be sufficient to withstand the con ression and bending stresses imposed while galloping. A single cannon bone has a greater resistance to bending than would four or five bones of the same total cross-sectional area. Hence a five-toed horse would require bones in its feet of greater total weight than a single-toed horse. Thus the single toe, like the concentration of muscles near the hip, reduces the weight of the foot, and consequently the energy needed for running.

This adaptation of a horse's leg for galloping has other consequences. Horses, unlike dogs, monkeys. and most human beings, cannot reach every part of the surface of their bodies either with their teeth or with one of their feet; they cannot scratch all over. Therefore they suffer from the attacks of blood-sucking insects. A horse's legs, therefore, are an example of 'specialization'. Although highly efficient in running, they have lost the capacity to perform another function, scratching, possessed by the legs of other mammals. This word 'specialized' is a difficult one, since it must apply to every animal or plant in greater or less degree. It should, however, be used only of an organ which, although efficient in one respect, lacks the capacity to perform other functions which are satisfactorily performed by similar organs in other animals.

This specialization of the legs of horses is associated with other adaptations. Horses are not completely defenceless against the attacks of insects. Their tails can be used to dislodge insects settling on their hind quarters, and their skin is loose and can be vibrated rapidly should an insect settle upon it. This picture of a major adaptation involving a number of secondary ones is common.

This discussion of the legs of horses is an example of what we may call the *a priori* method of recognizing adaptations. We argue from the function to be performed to the kind of organ likely to perform it effectively. Anyone familiar with animals and plants thinks like this a lot of the time, and it is a perfectly legitimate way to think. Most discussions, for example, of animal coloration are of this kind. Animals which live in Arctic regions where the ground is covered with snow are less easily seen if they are white, and many of them are in fact white. Most animals are darker above than below, and it is easy to show that in a world in which the light comes from above, objects so coloured are less easily seen than are uniformly shaded ones. The 'exceptions which prove the rule' are the water bugs which swim on their backs, and which have light-coloured backs and a dark ventral surface.

However, this method of argument has its dangers. Most important, the argument is seldom really *a priori*; we know what the legs of horses are like before wondering why it is efficient for them to be like that. There is therefore a constant danger of being illogical or fanciful in our explanations. Nevertheless, some adaptations are more elaborate and detailed than the most fanciful naturalist could have foreseen. Many moths and butterflies have a form and coloration which mimics that of dead leaves. The illusion may be heightened by the appearance of holes; in some cases there are genuine holes in the wings, in others a hole is suggested by the absence of scales over a part of the wing, and in still others by the shading and coloration of the wings.

Another danger of this *a priori* method is that not all differences between animals can be explained as adaptive. Like horses, kangaroos and ostriches inhabit open plains, and escape from their enemies by virtue of their high speed. Yet these three kinds of animals are very different in their methods of locomotion. Many of the structural peculiarities of each of these groups can be explained as adaptations to their particular modes of progression, but the reasons for the differences between them must be sought elsewhere. The only reasonable explanation lies in the differences in the structure and habits of the evolutionary ancestors of these animals.

It follows that any discussion of adaptation must ultimately rest on a direct study of the functioning of the organs of animals and plants. There is, however, another way of studying adaptation with which we shall particularly be concerned in this book. For example, in populations of the American field mouse *Peromyscus*, the coat colour is lighter in animals living in areas where the soil is sandy than where the soil is dark. Dice has been able to show in experimental conditions that owls take a larger proportion of mice whose colour differs from that of the background than of mice whose colour merges with the background. In this case, then, it has been shown that animals with a particular adaptation are in fact more likely to survive in a particular environment. This type of demonstration, where it is possible, is much more satisfactory than *a priori* argument; indeed, it is on such evidence, rather than on *a priori* argument, that the whole concept of adaptation rests. There are, however, two difficulties to be overcome. First, although it may be fairly easy to show differences in survival in laboratory conditions, it is much more difficult in wild populations; some cases where this difficulty has been overcome will be described in Chapter 13.

A second difficulty can be illustrated by returning to the legs of horses. Measurements show that the legs of a racehorse taper more than do those of a moorland pony, having relatively a stouter femur and slenderer cannon bone. However, this does not prove that the greater speed of the racehorse was due to this measured difference; in fact there must be many causes contributing to the difference in speed. Thus where there are many differences between two animals, it is difficult to disentangle the effects of these differences on capacity to survive in various environments.

Not all adaptations concern the structure or colour of animals. One example will be given of an adaptation of behaviour. The storks which breed in Europe winter in Africa. Storks breeding in western Europe start their migration in autumn in a south-westerly direction, thus avoiding the Alps, and travelling through France and Spain to cross into Africa by the straits of Gibraltar. Those breeding in eastern Europe start in a south-easterly direction, and reach Africa round the eastern end of the Mediterranean. It is known that the difference is inherited, not learnt from other birds. Nestling storks from east Prussia were taken to Western Germany, and released in autumn after the local birds had departed. These birds started off in a south-easterly direction, and were reported from the Alps and from Italy.

So far, animals and plants have been discussed as if they had a fixed structure, colour, behaviour, and so on which adapts them to particular conditions. This is only partly true, as can be seen by considering protective coloration in animals. Mammals and birds cannot change colour between moults, although many northern species, like the ermine, are white in winter and brown in summer. Many fish and amphibia and some reptiles, however, can change colour according to the background on which they live. Sometimes such changes are very rapid.

Flatfish, for example, can change both their colour and pattern in a few minutes; such changes are produced by the movement of minute granules of pigment within the cells of the epidermis. When the pigment is concentrated at the centre of the cell the animal looks pale, and when it is spread throughout the cell the animal looks dark. Still more rapid changes in colour are possible to squids and octopuses; pigment is contained in small bags which, being elastic, take up a spherical form, in which condition they cover only a small part of the surface, leaving the animal pale in colour. However, each bag can be flattened into the shape of a disc by a series of radially arranged muscles, each supplied by a nerve fibre. When flattened, the bags of pigment cover a large part of the surface, so that the animal appears dark in colour. By transmitting nerve impulses to the muscles, an animal can change colour in a few seconds. Colour changes in flatfish or in octopuses occur at a rate typical for physiological changes, such as the increase in the rate of heart beat after exercise, or the whitening of a man's skin when he is angry.

Other changes in coloration are much slower. Trout from a stream shadowed by trees are darker in colour than those from a shallow stream in the open. A trout transferred from one habitat to the other does not immediately change colour. The differences in colour are due to differences in the number of pigment cells and in the amount of pigment they contain. Here the changes resemble those of development and growth rather than those of physiology.

In fact, the difference between the mechanisms of colour change in flatfish and in trout is not as absolute as the preceding account suggests. Trout can change colour slightly by the migration of pigment, as do flatfish, and the underside of flatfish, which is normally pale, will gradually darken if they are kept in a glass aquarium illuminated from below, by the multiplication of pigment cells on the illuminated surface. Nevertheless, the distinction between rapid physiological changes and gradual developmental ones is valid, even though it is difficult to find examples of animals which can do only one or only the other.

Now the pale colour of a mouse living on a pale soil, of a sole lying on a sandy bottom, or of a trout in an open stream are all adaptations. But to use the same term for three types of coloration brought about in such different ways is liable to cause confusion, particularly when discussing evolution. It is therefore desirable to use different words for these three kinds of adaptation. Unfortunately there are no words for them generally accepted by biologists, although the distinction has long been recognized. In this book I shall use the terms 'genetically adapted', 'physiologically versatile or tolerant', and 'developmentally flexible'. These terms are, I hope, self-explanatory, but some examples may help to illustrate the importance of these different kinds of adaptation.

1. An animal or plant is genetically adapted to particular conditions if it possesses characters suiting it for life in those conditions, and if it develops those characters in all or most environments in which it is able to develop at all. Thus many birds and mammals are genetically adapted in that their colours render them less easily seen in their normal habitat, but their colour does not change if they are raised in unusual conditions.

2. Flatfish are physiologically versatile in being able rapidly to change their colour according to their background. The effect of such rapid physiological adjustment is to enable an organism to maintain its individuality in spite of changes in external conditions; in this case by avoiding being eaten by predators. A more usual kind of physiological adjustment results, however, not in a change in the external appearance of the animal, but in maintaining conditions within the body constant despite external changes. If you go into a cold room you may shiver, thus generating heat which keeps your body temperature constant, whereas on a hot day you sweat, and so are cooled down by the evaporation of water from your skin. This is as much an example of physiological versatility as is the colour change of flatfish. In both cases, an organism is able to survive in a wider range of conditions, because it can make appropriate and rapid adjustments to those conditions,

Animals vary greatly in the range of external conditions in which they can maintain themselves. For example, the concentration of salts in the body fluids of fish is lower than that in sea water, but higher than in fresh water. Consequently, most sea fish die in fresh water, because water enters through their permeable skin and gills until they become swollen, although they can survive in the sea by swallowing water and getting rid of excess salt through their gills. Freshwater fish can get rid of the water constantly entering their bodies by pumping it out through their kidneys and ureters, but die of water loss in the sea. However, a few fish, such as eels and salmon, can survive either in fresh or in saltwater, they are physiologically more tolerant.

Now all animals must be physiologically tolerant to some extent, because no animal lives in an absolutely unchanging environment. The nearest approach to this is at great depths in the sea. Fish inhabiting the open sea, where the temperature changes relatively little, cannot survive such large temperature fluctuations as can fish from rock pools, which may be heated in the sun and cooled at night.

Animals vary greatly in their tolerance of changes in their diet. Some mammals, like pigs and men, will eat almost anything. Horses are genetically adapted by the structure of their teeth and stomachs to feed on grass, a substance of little value as a food to ourselves, since our teeth would wear out if we attempted to chew it, and because it contains large quantities of cellulose which we cannot digest. However, the range of foods upon which horses can subsist is rather narrow, although this is no serious handicap so long as grass is as common as it is. The koala 'bear' of Australia is confined to a single food plant, the eucalyptus tree, a fact which severely limits its distribution.

3. An animal or plant is developmentally flexible if when it is raised in or transferred to new conditions, it changes in structure so that it is better fitted to survive in the new environment. The changes involved are gradual, and are usually brought about by cellular multiplication and differentiation.

This can be illustrated by various kinds of flexibility found in human beings. If a man does heavy manual work, the skin on the palms of his hands grows thick and horny. This change is induced by pressure on the skin, and is adaptive in preventing the outer layers of skin being worn away and the hands becoming sore. However, the skin on the soles of the feet becomes thickened whether or not it is exposed to pressure. Thus a man is genetically adapted in that he develops calluses on the soles of his feet, and is developmentally flexible in that calluses develop on the palms of his hands in response to pressure.

Another response to heavy work is a growth in the size and strength of the muscles. If for any reason the nerve supply to a muscle is lost, so that the muscle never contracts, the muscle atrophies; if a muscle contracts very frequently, it grows. In this way, a man's muscles. and to some extent his bones and tendons, become adapted by use so as to perform more efficiently the work habitual to them. Similarly, the legs of a horse are genetically adapted for running, but they are improved by use; a horse which is exercised gallops faster than one which is not.

There are many diseases, such as measles and chicken-pox, which a man usually catches only once. This is because, during an attack, for example, of chicken-pox, antibodies are formed in the body which can destroy the virus. Therefore one experience of the disease prepares the body to resist a second attack.

As a final example of such flexibility in man, consider the changes which occur when living at a considerable height above sea level. At great altitudes the supply of air, and so of oxygen, is reduced. So long as this reduction is not too great, it can be compensated by increasing the oxygen-carrying capacity of the blood. Oxygen is carried in the blood in chemical combination with haemoglobin in the red blood corpuscles. After living for some time at high altitudes, the number of these corpuscles is increased.

Not all changes which occur when an organism is exposed to new conditions can be regarded as examples of developmental flexibility. For example, human beings may suffer from rickets or from scurvy if their diet is lacking in particular vitamins. However, the changes involved do not enable them to survive better on a deficient diet, and may ultimately result in death. It is not always easy to distinguish between changes which adapt the organism to the conditions evoking them, changes which can therefore be regarded as examples of flexibility, and changes which indicate merely that the organism is unable to cope with the new conditions.

Adaptive changes in structure in response to changed conditions are commoner in higher plants than in higher animals. There is more room for such change; a man or a dog has a more or less fixed number of parts, for example, teeth, ribs, or fingers, whereas an oak tree may develop roots, leaves, and branches in varying numbers and forms, in accordance with the peculiarities of the soil in which it is growing or the incident light. Flexibility is often shown in the response of plants to light. A seedling must depend on the food substances which were present in the seed until it has grown green leaves in which, in the presence of light, new food substances can be synthesized. Seedlings kept in the dark usually grow more rapidly in height than those in the light. Ultimately, if kept in the dark, the seedlings die, but in natural conditions their more rapid growth increases their chances of reaching the light, and so surviving.

Many plant species are divided into races adapted to local conditions. These races are usually genetically adapted; they retain their typical form in changed conditions. The Russian botanist Michurin observed that first-generation hybrids between locally adapted races often showed the characters of the parent in whose habitat they were grown. For example, hybrids between broad-leaved cultivated pears and narrow-leaved wild pears from an arid region developed broad leaves if grown in an orchard, and narrow leaves in arid conditions. In this case the hybrids showed greater developmental flexibility than the parents. Clausen has recently obtained similar results with *Potentilla*, a relation of the strawberry. The lowland and subalpine races of *P. glandulosa* differ in height and in flowering time. The lowland form failed to survive in subalpine conditions, and the subalpine form was weak and stunted when grown in the lowlands. However, hybrids between the two races survived in either habitat, and resembled in height the parent in whose habitat they were grown.

The distinction between these different categories of adaptation, like most distinctions in biology, breaks down if it is pressed too far. Nevertheless it is desirable to make a distinction between genetic, physiological, and developmental adaptations, for the following reason. The fate of any population of animals will depend not only on how well adapted it is to the particular environment in which it lives, but also on how wide is its range of tolerance of changes in its environment. A species with a narrow range of tolerance is unlikely to spread to new areas, and unlikely to survive sudden changes in its environment. A species which can colonize new habitats is likely to have a longer evolutionary future; the first fishes which left the water to flop between pool and pool have inherited the earth.

It was suggested above that plants show greater flexibility in their patterns of growth than do animals; the latter, however, more than make up for the fixity of their structure by the developmental flexibility of their behaviour. When we speak of a man benefiting from experience, we usually have in mind changes in his behaviour. A man who has had a narrow escape when crossing a road learns to look both ways before stepping off the pavement. It is, however, worth pointing out that there is little difference in principle between a man who has learnt to look both ways, and one who, after one attack of chicken-pox, has 'learnt' how to resist further attacks. The difference is one of degree. The process of learning in its colloquial sense involves changes in the brain. We do not yet know the nature of these changes, but it is clear that the brain in higher animals has evolved a remarkable capacity to undergo adaptive modifications in the course of the lifetime of an individual, the result of these modifications being to alter the animal's behaviour in such a way as to increase its chance of survival.

The capacity to learn plays an important part in the success of birds and mammals. For example, Snow has recorded the nesting success of a number of individually ringed blackbirds in several successive years. He finds that the nesting success of birds breeding for the first time is lower than that of older birds, and lower than that of the same individuals a year later. This cannot be a mere matter of size and strength, since blackbirds, like the great majority of birds, are fully grown when they leave the nest. It is difficult to avoid the conclusion that they benefit by their experience. This may help to explain an otherwise puzzling fact.

Many large sea birds do not breed until their third or fourth summer. It is difficult to explain why it should be an advantage for a fully grown bird to delay its breeding, unless it is necessary for the bird to acquire sufficient experience to feed and protect its young.

We all benefit from the intelligence, and perhaps the good luck, of those of our ancestors who first learnt to control fire, or to domesticate animals and plants. But we are not alone in benefiting from the skill of our ancestors. For example, in many areas Great and Blue Tits have developed the habit of pecking through the lids of milk bottles standing on doorsteps to get at the cream. This is a habit which one tit will copy from another. It is not impossible that some individual tit Prometheus made the original discovery that there was a supply of fat under that unpromising covering, although the discovery may have been made many times. Similarly, many kinds of birds today nest on buildings. Many of them, like the swifts and the London pigeons, are descended from cliff-dwelling ancestors. The choice of a building rather than a cliff as a nesting site can hardly be genetically determined. More probably it is a habit which has spread because one bird has copied another, and because birds have nested in the same kind of place as that in which they themselves were raised.

These examples may seem rather trivial. However, they do demonstrate that the habits of a population may change far more rapidly than its genetic make-up. Further, those bird species which have been able to change their habits are the common English birds today; those which have not are confined to restricted areas, as is the Crested Tit to a few pine forests in Scotland, and the Bearded Tit to the fens of East Anglia.

CHAPTER 2

The Theory of Natural Selection

The fact that animals and plants are adapted to the environments in which they live was recognized long before the theory of evolution had gained general acceptance among biologists. Similarly, the idea that the different kinds of animals and plants could be classified according to a 'natural' scheme preceded the idea that such a scheme of classification reflected evolutionary relationships. It was in fact the similarities between different kinds, or species, of animals and plants, similarities which make a natural classification possible, which led Darwin, Lamarck, and others to seek an evolutionary explanation of the origin of species, just as it was the fact of adaptation which suggested to them theories as to how evolution might take place.

Before discussing the origins of Darwin's evolutionary theories, therefore, it is worth considering how far it is possible to speak of a natural scheme of classification in the absence of any such theory. The present method of classification can best be illustrated by giving the classification of a familiar animal, the lion. All lions are regarded as members of a single 'species', which is named *Felis leo*, such names always being written in italics. The inclusion of the word *Felis* in the name implies that lions belong to the 'genus' *Felis*, which also includes the leopard, *Felis pardus*, the tiger, *Felis tigris*, the Scottish wild cat, *Felis sylvestris*, and so on. All these cats have many things in common, including the possession of retractile claws. The genus *Felis* is included in the cat 'family', Felidae, comprising all the typical cats and also the cheetah, *Acinonyx jubatus*, which, although it resembles the other cats in most respects, has claws which are not fully retractile. The cat family in turn are included in an 'order', the Carnivora or flesh-eating mammals, an order which also includes the dogs, bears, weasels, hyenas, and others. The Carnivora are then grouped together with other animals which bear their young alive and suckle them in the 'class' Mammalia, which in turn are grouped with the fishes, amphibia, birds, and reptiles in the 'phylum' Vertebrata, or animals with backbones.

It will be seen that the classification is a hierarchical one. How far is it possible to regard it as a natural one, in the sense that the classification reflects real features of the variation of living things? Some of the difficulties which arise in the classification of animals and plants into different species will be described in Chapter 13; it will be argued that despite the difficulties the classification is natural in the above sense. For the present it is sufficient to point out that lions and tigers, for example, are sharply distinct from one another, and do not interbreed in wild conditions, although hybrids between them have been obtained in captivity, whereas, although not all lions are exactly alike, it would be difficult to divide the species into two parts, each of which could be regarded as a distinct species. Thus the classification of animals into separate species often corresponds to real differences, and is not an artificial scheme imposed for convenience; this is particularly true when considering only the animals and plants found in a particular region.

There is a sense in which the grouping together of different species into genera, families, and so on is an artificial procedure. It is of practical importance that every animal and plant should have a scientific name, and this requires that each should be placed in a genus; it will sometimes be a matter of opinion which genus should be chosen. For example, should the cheetah be placed in a genus of its own, *Acinonyx*, because its claws are non-retractile, or included in the genus *Felis* because in other respects it closely resembles the other cats? However, even though the decision on such questions may be determined by convenience or individual taste, the classification of animals into higher categories is not therefore wholly an arbitrary procedure. For example, it is generally agreed that the lion should be classified in the same genus as the tiger, leopard, and wild cat, and not in the same genus as the camel, although the latter classification could be supported on the grounds that the two animals are the same colour. Why should we base our classification on the common possession of retractile claws, and not on fawn coloration? The reason is not, as is sometimes thought, that resemblances of colour are in themselves trivial, and resemblances of structure fundamental. It is rather that a lion and camel have little in common except for their colour, and for the characters associated with their both being mammals, whereas the various kinds of cats resemble one another closely in the details of their limbs, backbones, skulls, teeth, viscera, and so on, differing only in coloration, size, and minor changes in proportions. Thus retractile claws are a better guide to classification than colour, because they are associated with a whole number of other characters, whereas animals which are the same colour may have little else in common. The recognition in particular cases of characters which are a valuable guide to classification depends on a study of the group in question. There are cases in which resemblances of colour are a better guide than of structure, For example, in the duck family, Anatidae, the presence or absence of a metallically coloured speculum in the wing, and the colour patterns of the downy young, are of value in classification. Of particular interest are the shovellers, a group of ducks with large spatulate bills. On the basis of a morphological character, the bill, the four species were classified into a genus, *Spatula*, distinct from the rest of the river ducks, *Anas*. However, further study of their behaviour, and in particular of their plumage patterns, suggests that the shoveller bill has been evolved several times, and that the various species of shovellers are in fact descended from different species of a particular group of river ducks, the Blue-winged Teal.

Classification is possible only because animals and plants do fall into groups resembling one another in a number of different respects. To take an example of a higher category than the genus, the amphibian order Anura (tailless) includes the frogs and toads. All members of the order resemble one another in their flattened skulls, short backbone, absence of tail, elongated hind limbs with an extra functional joint, in the structure of their hearts, and so on. Further, there are no living Amphibia which in these respects are intermediate between the Anura and other amphibian orders. However, some are wholly aquatic, some spend much of their time on land, and some live in trees. Thus structure rather than habitat is taken as a guide to classification; whales are classified as mammals because they have mammae and many other typical mammalian anatomical features, and not as fish, although they live in the sea.

Once the principle has been recognized that animals must be classified together because they resemble one another in a number of different ways, the classificatory scheme which is achieved will be much the same, whether the classifier believes that the species were created as variations on a number of themes, or whether he thinks that the resemblances between a group of species have arisen because those species are descended from a common ancestor. In fact the bases of our present classification were laid by pre-evolutionary taxonomists, of whom the most famous was Linnaeus. Linnaeus believed that species had been separately created, and that in devising his *Systema Naturae* he was uncovering the design of their creator. In this respect his philosophy resembled that of some physicists who have believed that in discovering fundamental laws of nature they were revealing the way in which God thinks.

A causal explanation, however, of the similarities between animals in terms of their relationships is possible once it is accepted that all living things are descended from one or a few kinds of simple living organisms. Such an explanation is particularly satisfying when two species of markedly different habits and external appearance are found to possess a fundamental similarity in structure and development, as, for example, do a whale and a monkey. These considerations led a number of biologists before Darwin's time to hold evolutionary views. However, at this level the theory of evolution may help to explain why a hierarchical classification of the animal and plant kingdom is possible, but it is no great assistance in improving such a classification in practice, nor does it suggest new lines of observation or experiment. Consequently, the fact of evolution was not generally accepted until a theory had been put forward to suggest how evolution had occurred, and in particular how organisms could become adapted to their environment; in the absence of such a theory, adaptation suggested design, and so implied a creator. It was this need which Darwin's theory of natural selection satisfied. He was able to show that adaptation to the environment was a necessary consequence of processes known to be going on in nature.

Before considering Darwin's ideas in detail, it is worth reviewing some of the other facts and ideas which influenced his thought. In addition to his knowledge of the problems of classification, including first-hand experience gained in his study of barnacles, two other fields of knowledge provided him with important materials for his theory. First, the results of artificial selection of domestic animals and plants revealed the enormous, but largely hidden, variability within a single species which could be made manifest by selecting and breeding from particular individuals. Second, during his voyages on H.M.S. *Beagle* he was struck by various features of the geographic distribution of animals which could most easily be explained by the hypothesis of evolution; an example will be discussed in some detail in Chapter 14.

He was also influenced by recent advances in the field of geology, and in particular by the work of his contemporary Lyell. Lyell's great achievement was to explain the past history of the rocks of the earth's crust in terms of processes such as erosion, sedimentation, and volcanic activity which can be observed at the present time. It was to be Darwin's role to explain organic evolution also in terms of contemporary processes. Thus the work of Lyell and his forerunners both provided Darwin with an account of the geological backcloth against which organic evolution has occurred, and set him an example in the methods whereby such evolution was to be explained. In fact, Darwin's first essay in evolutionary theorizing was in the field of geology, in his discussion of the origins of coral reefs, and not in the field of organic evolution.

The development of Darwin's ideas was also influenced by the social and economic conditions of his time. This is true not only in the direct sense that naturalists, particularly in England, France, and Holland, the possessors of colonial empires, were able to collect and study animals and plants from all over the world, and so were made aware of facts concerning geographical distribution and variation with an important bearing on evolution, but also by the more subtle processes whereby ideas derived from a study of social relationships influence the theories of natural scientists. Darwin was consciously influenced by the ideas expressed by Malthus in his *Essay on Population*. Malthus was concerned to justify the existence of poverty among a considerable section of the population; he argued that the human population is capable of increasing indefinitely in a geometric progression, and must therefore be held in check by the limited quantity of food available, and so by starvation. The argument is in part fallacious, since there is no evidence that the main factor limiting the human population is the shortage of food. However, the observation that animal and plant species, including the human species, are capable of indefinite increase in numbers in optimal conditions, is correct, and plays an important part in the theory of natural selection. Darwin must also have been influenced by the fact that he lived in the era of competitive capitalism, when some firms were improving their techniques, and increasing in size and affluence, while others were going bankrupt, and old crafts were dying out. It is unlikely that the concepts of competition and the struggle for existence in nature would have occurred to him so readily had he lived in a more static feudal society.

These various factors, the development of taxonomy, the study of domestication and of the geographical variation of living things, the development of an evolutionary theory of geology, and the concepts of competition derived from contemporary society, provided Darwin with the necessary methods of attack and materials for study; it required his individual genius to weld them into a comprehensive theory of organic evolution. His theory of natural selection starts from the observation that in optimal conditions, with unlimited supply of food and space, and in the absence of predators and disease, all animal and plant species are capable of increasing in numbers in each generation. In a few species, such as the herring, the maximum potential increase per generation may be as much as a millionfold. However, even in species such as our own where relatively few offspring can be produced by a single pair, the potential rate of increase is very rapid. If we assume, for example, that the average number of children, born to a married couple, who themselves grow up and marry is, in optimal conditions, only four, the population will double in each generation. The population would then increase a thousandfold in ten generations, and a millionfold in twenty generations, or about 600 years.

Since animal and plant numbers do not in fact increase indefinitely in this manner, it follows that either not all individuals born survive to sexual maturity, or that some sexually mature individuals do not breed, or that breeding individuals produce fewer offspring than they would under optimal conditions. At this point in the argument, a second fact based on observation is introduced; not all individuals in a species are alike. At least some of the differences between them will affect their chances of survival and their fertility. Some individuals will be better than others at catching food or escaping from predators, at finding mates or at raising their offspring. Just as a husbandman selects from his stock as parents of the next generation those individuals which seem to him best to meet his requirements, so in nature those individuals best fitted to survive in the given environment are selected as parents. This is the process of natural selection.

Now it is again a fact of observation that children tend to resemble their parents. In so far as this is true, the better adapted individuals in each generation, which survive and which leave most offspring, will tend to transmit to their progeny those characters by virtue of which they are adapted. Thus, by the combined processes of natural selection and of inheritance, the adaptation of the population to its environment is constantly perfected, or is constantly adjusted to a changing environment.

In later chapters some actual examples of natural selection based on observation of wild populations will be described. However, such cases are always complex, and it is difficult to collect all the relevant information. The process of natural selection will therefore first be illustrated by an imaginary numerical example, in which a number of simplifying assumptions have been made.

Let us suppose that in a population of mice in an area where the soil is dark, there are equal numbers of light- and of dark-coloured mice born in a given generation. We will follow the fate of 100 dark- and 100 light-coloured mice, counted at birth. It will be assumed that generations are separate, that is, there is a breeding season in the summer, and the animals which are parents during one summer have died before the next. This is not very far from the truth for small rodents; the assumption does not seriously alter the results of selection, but makes it much easier to think about numerically. We shall also assume that the numbers of males and of females which are born, and which survive to breed, are equal.

We will consider the effects of natural selection through predation by owls, which kill a larger proportion of the more conspicuous light-coloured mice. Suppose that 40 per cent of the light mice are killed by owls, and only 10 per cent of the dark ones. In addition many mice will die for other reasons, for example disease, cold, hunger, or predation by weasels, which hunt more by smell than by sight. Suppose the effect of these other mortality factors is to kill two-thirds of the mice which are not killed by owls; this mortality is unselective as far as colour is concerned, equal proportions of light and of dark mice being killed before the next breeding season. Thus the effect of predation by owls is to reduce the initial population of 100 dark and 100 light mice to 90 dark and 60 light mice, and of other mortality factors to reduce this population to a breeding population of 30 dark and 20 light mice. It does not matter in which order the various factors act. Thus the breeding population consists of 25 breeding pairs.

If the population is to maintain its numbers, these 25 pairs must produce 200 offspring, an average of 8 offspring per pair. Now the proportion of dark and of light mice among the 200 offspring will depend on how the colour difference is inherited. In the parental population there were 30 dark mice (60 per

cent) and 20 light mice (40 per cent). If, for example, mice
always resemble their mothers in colour, the proportions of the
two types among the 200 offspring would be approximately the
same as in the parental population, i.e. 120 dark and 80 light. It
is perhaps more realistic to suppose the light colour to be due to
a single Mendelian recessive (see next chapter); in this case, if
mating is at random, it can be shown that the numbers will be
approximately 116 dark mice (58 per cent) and 84 light mice
(42 per cent). In either case, the proportion of light-coloured
mice in the next generation is lower than in the original
population.

This example can be set out in the form of a table.

If selection continues to act in a similar manner in subsequent
generations, the proportion of light-coloured mice will continue
to fall, until finally such mice are very rare. Thus the effect of
selective predation by owls is to improve the adaptation of the
population to its environment.

This example will now be used to define the terms 'fitness'
and 'intensity of selection'. Of the original 100 dark mice
counted at birth, 30 survived to breed, and had an average of 8
offspring each.

Population of newborn mice	100 dark	100 light
Killed by owls	10	40
	——	——
Survivors	90 dark	60 light
$\frac{2}{3}$ mortality due to other factors	60	40
	——	——
Breeding population	30 dark	20 light

25 breeding pairs

↓

Average of 8 offspring per pair

200 newborn mice in the
next generation

If colour due to a single
  Mendelian factor

116 dark   84 light

Thus the 100 dark mice born had a total of 240 offspring, also
counted at birth, or an average of 2·4 offspring each, and the
100 light mice had 160 offspring: note that each offspring has

two parents, and so is counted twice, giving a total of 400. If a population consisting of equal numbers of the two sexes is exactly to reproduce its numbers in each generation, each individual must average two offspring. Thus the fitness of dark mice can be defined as 2·4/2, or 1·2; in the same way the fitness of light mice is 0·8, and the mean fitness of the population is 1·0.

Had all the original population of 200 mice been dark in colour, 60 individuals would have survived to breed instead of only 50, an increase of 20 per cent. This value of 20 per cent has been defined by Haldane[1] as the 'intensity of selection'; it gives a measure of how many lives are lost because not all individuals are as well adapted as are the fittest members of the population. Not all deaths will be selective in this way. Thus it is highly unlikely that owls would take only the light variety, and it was in fact assumed that some dark mice are killed by owls. There was also a two-thirds mortality due to other causes, which is unselective as far as colour is concerned. It is of course likely that there would be differences other than of colour between individuals, which would make some more likely to survive than others. The value of 20 per cent measures the intensity of selection only for colour differences due to predation by owls. The total intensity of selection acting on all differences between individuals would be higher.

One last point can be illustrated from this example. After a number of generations the population will come to consist almost entirely of dark individuals. Now the fitness of dark individuals was initially 1·2. However, the mean fitness of a population consisting wholly of dark mice could not remain indefinitely at this value, since that would imply an indefinite increase in numbers; the mean fitness of the population must fall again to unity. This may happen because owls continue to take the same total number of mice, but, because there are no longer any pale and conspicuous mice, are forced to search until

---

[1] Strictly, Haldane defines the intensity of selection as $I = \log_e 60/50 = 0\cdot183$. For small values of I this gives approximately the same value as the definition used above, but Haldane's definition has the advantage that, if selection is acting on a number of different characters independently, the total intensity of selection is equal to the sum of the intensities for each character acting by itself.

they can find dark ones. Alternatively, the population may be held in check by an increase in the mortality due to other causes, for example disease or food shortage.

An actual example of a change in a wild population which has happened because some individuals are more easily found by predators than others will be discussed in Chapter 10. For the rest of the present chapter the role of natural selection in determining the rate of reproduction will be discussed. It will be recalled that Darwin's theory started from the observation that the reproductive rate of all species in optimal conditions is greater than that necessary to maintain their numbers. This observation is correct, so that the ensuing discussion does not invalidate Darwin's argument. Yet there are enormous differences between the maximum potential rates of increase of different species, which are determined by natural selection. The evidence for this view has recently been discussed by Lack, on whose work the following account is based.

The number of eggs laid by different kinds of animals varies enormously. For example, herring and cod lay many millions of eggs, whereas sharks and rays lay relatively few. Herring lay minute eggs which hatch into larvae which form part of the plankton. The great majority are eaten by other planktonic animals before they grow and mature. Sharks and rays lay large yolky eggs which hatch into small fish six inches or more in length. The eggs are either protected by a horny egg-case, or retained within the mother's oviduct until they have completed their development into small fish. Thus a much larger proportion of eggs survive to become adult fish.

Now it could be argued that, in view of the high larval mortality, it is necessary for herring to lay large numbers of eggs if the species is to survive. This is true enough, but surely it would be more efficient (i.e. would confer a selective advantage) to lay a large number of eggs which were at the same time as well protected as those of a shark. As soon as the problem is posed in this way, the answer is obvious. The high survival rate of young sharks arises because the highly vulnerable free-living larval stage is avoided by storing large quantities of food in the form of yolk, in each egg. The total amount of yolk which a female can store in her eggs depends on the quantity of food which she herself can capture, over and above that required to keep herself going. There are therefore only the alternatives of laying a few large eggs or many small ones.

We do not in general know why different species have adopted, some one method, some the other, but it is sometimes possible to guess at the answer. For example, Kramer recorded the number of eggs per clutch laid by the wall lizard *Lacerta sicula* on the Italian mainland and on the offshore islands. His results were as follows:

	Mainland	Offshore Islands
Number of eggs per clutch	4–7	2–4
Size of eggs	small	large

A possible explanation is that on the mainland the chief mortality is due to predators, and a large number of eggs increases the chances that some offspring will survive. On the islands predators are rare, but food and water scarce. In such circumstances a larger supply of food and water in the eggs may increase the chances of survival.

In this example, although the facts are known, the explanation is no more than a guess. In plants, Salisbury has been able to show some of the factors which influence whether a few large seeds or many small ones are produced. He finds that the smallest seeds are produced by species in open habitats, such as fields and disturbed earth, and that successively larger seeds are produced by species inhabiting scrub and woodland margins, by the herbaceous flora of woodlands, and by woodland shrubs and trees. Thus the larger seeds are produced in habitats where the seedling must grow to an appreciable height before reaching the light, until which time it must rely on the food reserves present in the seed. This interpretation is confirmed by the fact that the seeds of desert and dune plants are usually large, since the seedling must strike deep roots before reaching moisture.

This discussion of the relation between seed size and habitat is an example of the *a priori* method of analysing adaptations. There is, however, some direct evidence of the effects of natural selection on reproductive rate in birds. The number of eggs laid by a bird is not limited by the supply of food to the hen, since, if eggs are removed from the nest during the laying period, the hen will continue to lay eggs until the number is made up. Lack argues that 'clutch size has been adapted by natural selection to correspond with the maximum number of offspring for which the parents can, on average, find enough food'.

The English Swift, *Apus apus*, usually lays either two or three eggs. Nestling swifts can survive short periods of starvation, but long periods nevertheless cause death. Death from starvation is commoner in broods of three, as is demonstrated by the following figures, collected during 1946–50:

Brood size	% Fledging	Mean number Fledging per nest
2	82	1·64
3	45	1·35

Thus females which lay 2 eggs leave more progeny than those which lay 3 eggs, although the latter might be at an advantage in particularly favourable years. Both kinds of female leave more progeny than would females laying only 1 egg. It follows that there is an optimal brood size favoured by natural selection.

In most other passerine birds, the percentage of nestlings which fledge does not vary with the size of the clutch, so that natural selection cannot be determining clutch size in quite the same way. However, the weight of fledglings from large broods is lower than that of fledglings from small broods. This suggests that the subsequent chances of survival of birds from large broods may be lower. Unfortunately, once birds have left the nest it is impossible to follow all of them individually. It has been possible to get round this difficulty in the case of Swiss Starlings. A number of birds were ringed while still in the nest, and records kept of the size of the clutch from which they came. In autumn, the starlings migrate to North Africa. Thus, if a ring was returned from North Africa, it showed that the bird had successfully completed its first migration. In this way, it was possible to compare the chances of survival of birds from broods of different sizes, and by multiplying these chances by the numbers in the brood, to compare the average number of survivors from broods of different sizes. It was found that the number of survivors per brood increased with brood size up to a size of 5 eggs per clutch, but that for clutches of more than 5 eggs the increased number of birds fledging was counter-balanced by the increasing mortality during the first migration. The value of 5 eggs per clutch corresponds to that most commonly found.

Thus for both swifts and starlings, it has been shown that there is an optimal size of clutch, which gives the greatest total number of survivors, and that this optimal clutch size is also the one most commonly found in the population. Some birds lay more and some less than the optimal number; in either case they leave fewer progeny than do birds laying the optimal number. There is, however, one link in the chain of argument still missing. We do not know how these differences are inherited, or if they are inherited at all. The selective mortality which has been demonstrated will only be effective in preserving the adaptation of the population if female birds tend, on the average, to lay the same number of eggs as their mothers. This is probably true, but a discussion of examples of natural selection in which the inheritance of the differences is also known will be postponed until the laws of heredity have been described.

Most, though not quite all, examples of natural selection which have been studied resemble the ones just described in that selection acts so as to maintain, rather than to change, the adaptations of a population. This is to be expected, except where a population has recently colonized a new environment, or is living in a rapidly changing one. Natural selection may, however, act so as to adapt a species in different ways in different parts of its range. In many kinds of birds the average number of eggs in a clutch is low nearer the tropics and increases in more northern latitudes. For example, in the Canaries the average clutch size of robins is 3·5, it is 5 in southern England, and over 6 in Scandinavia. This is probably because the length of daylight, and hence the time during which the parent birds can search for food, increases with latitude.

The number of eggs per clutch may also vary from year to year in the same region. The clutch size of owls and hawks is greater in years when the voles upon which they feed are abundant. Similarly, many species from arid regions in Africa and Australia lay fewer eggs in particularly dry years. In both these cases the number of eggs laid may be determined by the supply of food at the time of laying. More remarkable is the observation of Gibb that the mean clutch size of Great Tits in the same wood varied from 8 to 12 in different years, and was greatest in years when the caterpillars on which tits feed their young were commonest. Now the first brood is laid when caterpillars are still scarce, although by the time the young birds hatch they are abundant, whereas the second brood, which is smaller than the first, is laid when the caterpillars are at their peak. It follows that the clutch size cannot be dependent on the supply of food at the time of laying, presumably the same conditions which favour an abundance of caterpillars in a particular year also influence the tits to lay a larger number of eggs.[2] In these cases, the number of eggs which a bird lays is not fixed by its genetic make-up, but is influenced by environmental conditions; nevertheless the flexibility of behaviour which enables birds to adjust the number of eggs they lay to the supply of food may be inherited, and may be as much the result of natural selection as are the genetic adaptations of clutch size to food supply described in swifts and starlings.

[2] In fact the most important environmental factor is the temperature during March.

CHAPTER 3

Heredity

The effectiveness of natural selection in improving the adaptation of a population to its environment depends on how far the differences between individuals which are responsible for their success or failure in the struggle for existence are inherited by their offspring. Great advances in our understanding of this problem have been made since Darwin's time, starting with Mendel's discovery of the laws of segregation, followed by the detailed studies of inheritance in the fruitfly *Drosophila* by Morgan and his colleagues, and more recently by the unravelling of the molecular basis of inheritance. Ideas on how evolution occurs have been much influenced by these advances. Therefore some of the salient points will be summarized in this chapter.

There are two methods of approach to problems of heredity. The first is to observe what is in fact transmitted directly from parents to their offspring; that is, to observe the structure and development of the sex cells, or 'gametes', produced by the parents, since it is the union of two gametes, the egg and sperm, which forms the starting point of the next generation. Therefore the physical basis of heredity must be contained in these gametes. In the case of mammals and of other viviparous animals this approach also requires a study of substances acquired by the young from its mother, either through the placenta or in the mother's milk.

The second method of approach is to study sets of related animals, for example parents and their offspring, or brothers and sisters. In so far as all the individuals in a family resemble one another, little can be deduced except an affirmation of the principle that 'like produces like'. Consequently such investigations usually start from a cross between individuals which differ to a more or less marked degree, or from the observation that one or more of the offspring from a cross differ from their parents.

The former of these two approaches is the field of the cytologist, the latter of the geneticist. The two branches of science are closely connected; it is always the aim to interpret the results of the one in the light of the other. It is therefore convenient to start a discussion of heredity by describing a case where the tie-up between genetics and cytology is well understood. This is so in cases of simple Mendelian inheritance, so called because such phenomena were first described by Mendel in peas. In the course of the description it will be necessary to introduce a number of technical terms which may be new to the reader.

As an example, consider the inheritance of the character 'dumpy wings' in the fruitfly *Drosophila melanogaster*. In dumpy flies the wings are short, with blunt tips, in contrast to the normal elongated wings with smoothly rounded tips. If two dumpy flies are crossed, the offspring are all dumpy; the character 'breeds true'. When a dumpy fly is mated to a normal one, the first generation, or F_1, are all normal. However, if two of these F_1 flies are mated together, in the second generation, or F_2, both normal and dumpy flies are obtained; the character is said to 'segregate' in the second generation, If the numbers of normal and of dumpy flies in the second generation are counted, it will be found that there are approximately three times as many normals as dumpies; this is the Mendelian 3:1 ratio. The dumpy flies from the F_2 breed true, as did the original dumpy flies. However, if the normal flies from the F_2 are crossed to dumpy flies, it will be found that they do not all behave alike. Approximately one-third of the normal F_2 flies, mated to dumpies, will give nothing but normal offspring; they resemble in this respect the original normal flies. The other two-thirds, mated to dumpies, give normal and dumpy offspring in approximately equal numbers.

These results are summarized in Figure 1, under the heading

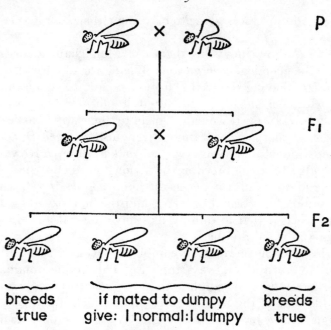

P

F₁

F₂

breeds
true

if mated to dumpy
give: I normal:I dumpy

breeds
true

Figure 1. The geneticist's findings.

'the geneticist's findings'. In Figure 2 is shown the explanation of these facts suggested by Mendel. Each gamete, egg or sperm, is assumed to contain a single factor, which may be either D for normal wings or d for dumpy wings. An individual, or 'zygote' formed by the union of two gametes, therefore contains two such factors, and may be either D/D, D/d, or d/d. Individuals which contain at least one D factor $(D/D, D/d)$ have normal wings; those which contain no D (d/d) have dumpy wings. This is expressed by saying that the factor for normal wings is 'dominant' over the factor for dumpy wings, or conversely that dumpy wings are 'recessive' to normal ones. It is conventional to use a capital letter for the dominant factor, and a small letter for the recessive one.

An individual with two similar factors $(D/D, d/d)$ is referred to as a 'homozygote', or like-zygote, and can only produce one kind of gamete, carrying the factor in question. A D/d individual, carrying two unlike factors, is called a 'heterozygote', and produces gametes carrying either D or d in equal numbers.

Figure 2 shows how this theory of Mendelian factors can explain the genetical observations. The F_1 are all normal, since all are D/d, having received D from one parent and d from the other. However, in the F_2 individuals are formed in the ratio 1 D/D:2 D/d: 1 d/d. If we judge solely by their appearance, i.e. by their 'phenotype', we cannot distinguish between D/D and D/d individuals, and so observe 3 normals: 1 dumpy. However, if we judge by their breeding behaviour, or 'genotype', it is possible to distinguish between the homozygotes, D/D, which give nothing but normal progeny, and the heterozygotes, D/d which, when mated to dumpy flies, give normal and dumpy offspring in equal numbers.

In this example one factor is completely recessive to the other. This is by no means always the case. The simple dominant-recessive relationship implies that the heterozygote exactly resembles one of the homozygotes, from which it can only be distinguished by breeding tests. However, the heterozygote may be intermediate between the two homozygotes, or may be qualitatively different from either of them. The importance of this fact will emerge when discussing some examples of artificial and of natural selection.

The essential features of the Mendelian theory are:

(*a*) Only one factor of a pair (in the example, the pair concerned with the difference between normal and dumpy wings) is present in a gamete, and two are present in an individual developing from a fertilized egg.

(*b*) In a heterozygote, D/d, the unlike factors do not blend or merge with one another, but are transmitted to the gametes which go to form the next generation with the same properties as they possessed when they entered the individual at fertilization. This is true also in cases in which the phenotype of the heterozygote is intermediate between the two homozygotes; the effects of two different factors may interact in various ways during the development of a heterozygous individual, but the factors themselves are transmitted to the gametes in their original form.

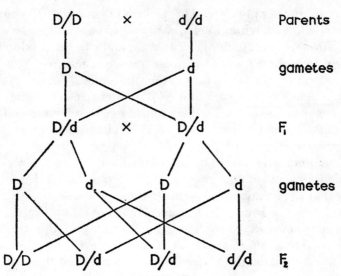

Figure 2. Mendel's theory of factors.

The following question can now be asked: Are there any visible structures of which two are present in a zygote, but only one in a gamete? In fact the chromosomes, which are stainable threads visible in the nuclei of cells during cell division, answer to this description. In the fertilized egg of *Drosophila melanogaster* there are 8 such chromosomes. They do not all look alike; there are 4 pairs of chromosomes, the two members of a pair resembling one another very closely, and differing from members of other pairs. The 4 kinds of chromosomes have been given numbers, I (rod-shaped), II and III (V-shaped), and IV (dot-shaped). A pair of similar chromosomes are said to be 'homologous' to one another.

At each cell division during the development of egg into adult, the chromosomes also are reproduced, so that each cell in the adult body resembles the fertilized egg in having two similar sets of 4 chromosomes; a nucleus having two sets of chromosomes is 'diploid'. (In some tissues the cells may have numbers of chromosomes other than 8, but this does not seriously affect the argument.) However, in the production of gametes a single cell with 8 chromosomes divides twice, while the chromosomes are reproduced only once. This process of 'meiosis' (lessening) gives rise to 4 daughter nuclei, each containing only a single set of 4 chromosomes; such nuclei are 'haploid', In the production of eggs, 3 of these nuclei, the polar bodies, degenerate, and the 4th becomes the nucleus of the unfertilized egg. In the production of sperm, each of the 4 daughter nuclei becomes the nucleus of a sperm, At fertilization the egg and sperm nuclei fuse, each providing one of the two sets of chromosomes of the new zygote.

This process is shown diagrammatically in Figure 3. Only one of the four pairs of chromosomes has been included. It will be seen that the behaviour of a chromosome is identical with that postulated for Mendel's factors. However, the 'factor' for dumpy wings does not correspond to a whole chromosome, but to a particular short region of chromosome II. Such a short region is conveniently referred to as a 'gene', and the position along the chromosome which it occupies as a 'locus'. To put the matter in a different way, there is a particular region or locus of chromosome II in *Drosophilia* which can exist in two different forms, which have different biochemical actions during development, and which are symbolized by the letters D and d. When a given region of a chromosome may be present in biochemically different forms in different individuals, or in the two homologous chromosomes in the nuclei of a single individual, the different forms are referred to as different 'alleles'. It is clear that whereas two alleles at a given locus can occur together in a zygote, only one can be present in a gamete.

In Figure 3, the allele d has been represented by a white band, the allele D by a black band on the chromosome. It must be emphasized that the two kinds of chromosomes cannot in fact be distinguished under the microscope. However, in a few cases of Mendelian inheritance, for example of notched wings in *Drosophila*, we are concerned not with two biochemically different alleles, but with the complete absence of a short region of a chromosome as compared to its presence. In such cases it is possible to distinguish the two types of chromosome under the microscope, not, it is true, in the gametes, but in the giant chromosomes in the salivary glands.

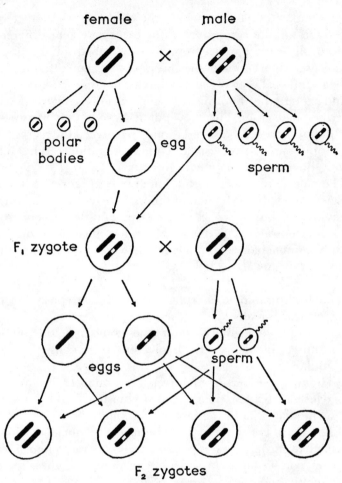

Figure 3. The cytological explanation. The gene for dumpy wings is shown as a white band across the chromosome, although in practice it cannot be distinguished under the microscope from its wild-type allele.

The main point which the preceding argument demonstrates is that there is a close agreement between the observed behaviour of chromosomes and the segregation of the character 'dumpy' in sexual crosses. That this correspondence reflects a real causal relationship, and is not purely fortuitous, is confirmed by studies on abnormal types of inheritance. There are a number of cases in which the mode of inheritance of a character, although apparently due to a single gene difference, shows a different pattern from that described for dumpy. In many cases observation of the chromosomes shows that they too have an unusual type of segregation at meiosis, of a kind which would explain the abnormal inheritance. This is not the place to describe such cases, but they put it beyond any reasonable doubt that there is a causal relationship between the segregation of chromosomes at meiosis and the segregation of characters among the progeny of sexual crosses.

The picture of gamete formation given in Figure 3, although sufficient to explain Mendelian inheritance, cannot explain the behaviour of 'linked' genes–i.e. of genes at different loci on the same chromosome. The relevant facts are illustrated in Figure 4. Only a single pair of homologous chromosomes is shown. The important events are as follows:

(a) Each chromosome is replicated, so that it consists of two identical threads.
(b) Pairs of homologous chromosomes come to lie side by side, forming 'bivalents'. Each bivalent then consists of four similar threads.
(c) The two members of a pair repel one another, but are held together at a few points, called chiasmata; there are two such chiasmata in the bivalent in Figure 4.
(d) There are two successive divisions of the nucleus, without further chromosome replication, giving rise to four nuclei, each containing a single set of chromosomes; these are the gametic nuclei.

In the figure, the chromosome derived from the father (paternal) is shown cross-hatched, to distinguish it from the maternal chromosome. In the four-strand and later stages, sections of chromosome which are copies of the original paternal chromosome are likewise shown cross-hatched. Now it is not normally possible to distinguish maternal and paternal chromosomes under the microscope. Consequently the conclusions, incorporated in the figure, concerning the paternal and

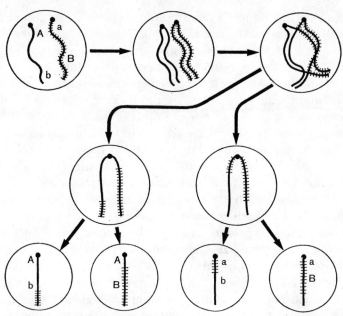

Figure 4. The behaviour of a pair of 'homologous' chromosomes during meiosis. The original paternal chromosome, and segments of chromosomes which are copies of it, are shown cross-hatched, although in practice there is no visible difference between a pair of homologues.

maternal contributions to the chromosomes of the gametes (i.e. that each consists in part of segments copied from the original paternal, and in part from the maternal chromosome), are based on genetic experiments and not on direct observation of the chromosomes. It is, however, clear that there is a connection between the formation of chiasmata and the recombination of maternal and paternal elements in a single thread. At present rapid progress is being made towards understanding the process of recombination in molecular terms.

The genetic consequence of all this is that if two genes are inherited from the same parent, they tend to be transmitted together, but if a recombination takes place between them, one is transmitted without the other.

There is one other genetic effect of chromosomes to be described, namely their role in the determination of sex. In most animals sex is genetically determined, in the sense that the sex of an individual is fixed when the zygote is formed at fertilization, although it is sometimes possible to reverse the sex of an individual by appropriate experimental procedure. It is therefore natural to look for a difference between the chromosomes of males and females, and in many species of animals such differences can be found. For example, it was stated above that in *Drosophila melanogaster* there are two sets of four chromosomes, the members of one set exactly resembling those of the other in appearance. In fact this is true only of female *Drosophila*. In males there are two pairs of V-shaped chromosomes and two dot chromosomes, but, in place of the pair of rod-shaped chromosomes of females, there is one rod-shaped chromosome and one J-shaped chromosome in each nucleus. The rod-shaped chromosomes are called X chromosomes, the J-shaped ones Y chromosomes. As far as this pair of chromosomes is concerned, a male is XY and a female XX. At meiosis, females produce eggs all of which contain a single rod-shaped X chromosome, but males produce two kinds of sperm in equal numbers, one kind carrying an X chromosome and the other a Y. The sex of an individual is then determined at fertilization by the kind of sperm which penetrates the egg; an egg fertilized by an X-bearing sperm develops into a female, and by a Y-bearing sperm into a male. A similar method of sex determination is found in most mammals, including man; in birds and in moths there is also a visible difference between the chromosomes of the two sexes, but it is the female which has two unlike chromosomes, i.e. is XY and the male two like chromosomes, XX. In fish and amphibia there are no visible differences between the chromosome complements of the two sexes. However, genetical experiments have shown that in these groups, as in *Drosophila* and in man, usually the males are genetically heterozygous and the females homozygous, but the differences between the two sex chromosomes of a male, although effective at a biochemical level in determining sex, cannot be detected under the microscope; in this respect they resemble the allele differences which produce either dumpy or normal winged flies.

Many differences between animals and plants are inherited

in the same manner as dumpy wings in *Drosophila*. In mice, for example, this is true of albinism as opposed to colour, the unbanded hairs of black mice as opposed to the banded hairs of wild mice, the habit of waltzing, and so on. The range of characteristics whose inheritance can be explained by the Mendelian mechanism can be further extended, once two complications are recognized: several characteristics may be influenced by a single gene, and a single characteristic may be influenced by genes at many loci. These two complications will be discussed briefly, before returning to the more important topic of the nature of genes or Mendelian factors.

(i) *The Manifold Effect of a Single Gene*

When discussing the inheritance of dumpy wings, it was assumed that a change in a single gene resulted in a change in a single character, the wings. This is a serious oversimplification. For example, mice homozygous for the gene 'grey lethal' have a grey coat and usually die at weaning because their incisor teeth fail to erupt through the jaws. A study of their skeletons shows that every bone in their bodies is unusual in shape. Gruneberg has been able to show that all the skeletal changes are due to a single cause. In normal mice, bone is not only laid down during development, but in other places it is resorbed; thus many bones are hollow, because as new bone is laid down at the surface during growth, bone is resorbed at the centre. In grey lethal mice, bone is laid down normally but it is not resorbed. Consequently the cavities of the limb bones are not hollow, but contain many spicules of bone. This also explains why the teeth do not erupt; in normal mice, holes appear in the surface of the jaws through which teeth then grow. In grey lethal mice, no such holes appear.

It has not been possible to explain the connection between the changes in the skeleton and the grey coat colour. Nevertheless it is thought that all the effects of a single gene are due to a single change in the biochemical activity of that gene. Our failure to explain the connection between skeleton and colour illustrates our ignorance of how the biochemical activities of genes influence the development of animals.

The importance of such observations is this. If a single gene change results in changes in two or more apparently unconnected characters, for example teeth and coat colour and hollow bones, it suggests that these characters are not as unconnected as they seem. It is then worth while to seek for causal connections in development between them. Such causal connections in development could be important in evolution. For example, it is known that in many horned mammals growth in size is associated with a relative increase in the size of the horns. Therefore natural selection for increased size might also carry with it a relative increase in horn size, even though the latter did not confer any advantage. However, this argument must not be pushed too far. If the increase in the size of the horns was a serious disadvantage, we would expect that natural selection would alter the association between the size of the animal and of its horns, so that increased size would no longer involve larger horns.

A single gene change may affect several different characters; conversely, the same characters may be influenced in the same way by changes in several genes at different loci. For example, in *Drosophila subobscura* there are three genes, hoary, frosty, and rimy, all on different chromosomes, but all producing identical phenotypic effects. There are white hairs growing out between the facets of the eyes, giving a frosted effect, and the wings are crumpled in a characteristic way. Only one similar gene is known in all the other species of *Drosophila*. These facts suggest the following conclusion, first pointed out by Spurway. The development of *D. subobscura* is such that a simple biochemical change in the fertilized egg can alter development so as to produce both frosted eyes and crumpled wings; the absence of similar phenotypes in other species would be explained if in their eggs the same biochemical changes caused a breakdown of development and death. To put the matter in another way, the pattern of development of a given species is such that there are only a limited number of ways in which it can be altered without causing complete breakdown. The fact that the same phenotypic abnormalities can often be produced either by a genetic change or by environmental conditions, as in the case of dumpy wings, also supports this idea.

Now the more closely related are two species of animals, the more similar will be their patterns of development, and so the more likely is it that a simple genetic change will produce similar changes in the character of the two species. This helps to explain a phenomenon noted by Darwin, that of 'analogous variation'. If two groups of animals or plants are closely related to one another, they tend to resemble one another not only in their typical appearance and structure, but also in the kinds of variation found among them. For example, cats and rabbits resemble one another not only in a number of anatomical features, but also in their colour varieties; there are black cats and black rabbits, Siamese cats and 'Himalayan' rabbits, cats and rabbits with patches of white fur, and so on. Now if it is true that in one group of animals the patterns of development make possible a wide range of variation in a particular character, whereas in another such variation is difficult, we should expect to find that evolution has led to great differences between species in one group of animals in a character which is relatively constant in another. An example will make this point clearer. A diagnostic feature of mammals is that they possess double-rooted molar teeth, but a far more striking feature of their dentition is that the teeth in different parts of the jaws differ in shape and function. For example cats have small incisors, large stabbing canines, and scissor-like molars, whereas horses have cropping incisors, no canines, and high ridged grinding molars. In contrast to this variation in the structure of the teeth from species to species, and from one part of the jaws to another in mammals, reptiles tend to have relatively simple teeth whose shape varies little from the front to the back of the jaw. One feature of the development of mammals which may have favoured the evolution of their highly specialized teeth is that the teeth are replaced only once during a lifetime, whereas in reptiles there is continuous tooth replacement throughout life. If teeth are continually falling out and being replaced by others, it is difficult to maintain the accurate fit between the teeth in the upper and lower jaws which is necessary if the specialized teeth of mammals are to work efficiently; for example, the carpassial teeth of carnivorous mammals, like the scissors they resemble, will cut only if there is a proper relationship between the two blades. In other words, a pattern of tooth development common to all mammals has favoured the evolution of enormous variation in detailed structure between them. This example helps to show how important it is in discussing evolution to remember that, whereas differences between animals may be transmitted in sexual reproduction by differences between chromosomes, these chromosomal differences produce their effects during individual development.

(ii) *Multifactorial Inheritance*

To return again to the inheritance of dumpy wings, it will be recalled that individuals fall into one of two sharply defined classes dumpy or normal, with no intermediates. Now there are many characters for which such a classification would be impossible. For example, it is certain that at least some of the differences in stature between human beings are genetically determined, but it would be impossible to classify people as either 'tall' or 'short', except by singling out the small number of very short people, or 'dwarfs'. Figure 5 shows the frequency of men of different heights. Most men are of medium height, and very tall or very short men are rare. Much of the variation in animal and plant population is like this.

In the early days after the rediscovery of Mendel's laws it was widely held that the inheritance of such continuously varying characters could not be explained in terms of Mendelian factors, and that some quite different kind of mechanism must be invoked. However, it was soon realized that if stature in man is influenced, not by a single pair of allelic genes, but by many pairs of alleles at many different loci, each allele difference by itself producing only a small effect on stature, then distributions similar to that in Figure 5 would be expected. The fact that the

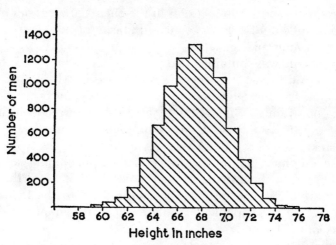

Figure 5. Distribution of stature in 8,585 adult men born in Britain. There were two individuals between 57 and 58 inches, and two between 77 and 78 inches, but the scale of the diagram is too small to show these.

genes themselves behave as discrete units does not mean that the characters they influence, for example stature, must fall into a few discrete classes. Provided that there are enough different genes influencing a given character, that character is likely to vary continuously. Experimental studies of this kind of inheritance will be described in Chapter 9.

We must now return to consider the nature of genes. The Mendelian pattern of inheritance requires that the factors or genes have the following remarkable properties:

(i) A vast number of different kinds of genes must be possible.
(ii) A process of exact replication or copying must occur, so that when a cell divides identical sets of genes should be passed to each daughter cell.
(iii) Genes must in some way influence development.

For many years, the major puzzle in genetics was the apparent contradiction between the capacity for exact replication, which suggests a degree of independence from changes within the cell, and the capacity to influence development, which requires active intervention in the chemical activities of cells. This difficulty has in large measure been solved by recent advances in molecular genetics.

The first clue was contained in a paper published by Griffith in 1928. Griffith was studying the pneumococci which cause pneumonia. He injected into a mouse two strains of pneumococcus; one strain was living, but of a non-virulent type which was not expected to cause disease symptoms; the other strain had been killed by heat, but was of a type which if living would cause pneumonia. The injected mice died of pneumonia, and Griffith found in them living virulent bacteria. It seemed that something had been transferred from the dead to the living bacteria which was capable of giving to the latter an inherited property of the former.

The significance of this discovery was not understood at the time, and did not become apparent until the publication in 1944 of the work of Avery and his colleagues. They showed not only that Griffith's results were correct but that characters other than virulence could be transferred in this way, and that transference did not only occur in a mouse, but would take place if living and dead bacteria were mixed together in a test tube. They then settled down to discover which chemical component of the dead bacteria was responsible for the transformation. Killed and smashed up bacteria were treated in various ways, so as to destroy various components, and the transforming ability of the various extracts were then determined. It was found that the transforming ability resided in the deoxyribonucleic acid component, or DNA.

The discovery came as a considerable surprise. This may seem odd, since it was known that chromosomes are composed of DNA and protein, and that DNA is found in few other places in the cell, whereas proteins are ubiquitous. DNA therefore seemed a natural candidate as the material of which genes are made, and therefore as a hereditary transforming factor in bacteria. The reason for the surprise was that an erroneous idea was held as to the structure of DNA. It was known that DNA is a long thread-like molecule, built up from four kinds of smaller molecules, the so-called bases, adenine, thymine, guanine, and

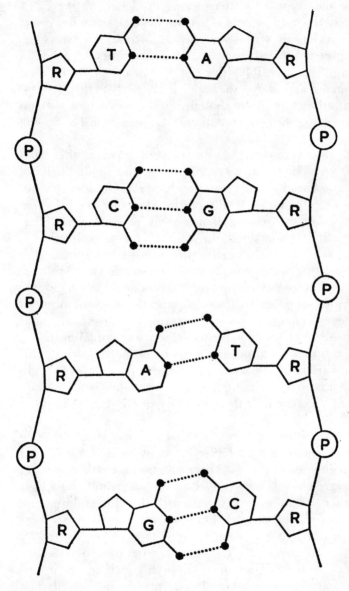

Figure 6. The structure of DNA. The letters A, T, C, and G represent the four bases adenine, thymine, guanine, and cytosine respectively; R and P represent the ribose sugar and the phosphate forming the backbone of each strand. The dotted lines represent the bonds linking complementary bases together.

cytosine. It was thought that these four bases were arranged in a regular and repeated pattern along the length of the DNA molecule, so that any piece of DNA was just like any other piece, except for its length. If this were so, DNA could not satisfy the first requirement listed above, that many different kinds of genes should be possible, since it is hardly plausible that different genes should owe their specificity solely to their length. DNA had therefore been thought to be a kind of skeletal material holding the genes together, the genes themselves being proteins.

Avery's discovery demanded a re-examination of the structure of DNA. The result was the now famous double helix of Watson and Crick, based on the X-ray crystallography of Wilkins. The chemical details of this structure do not here concern us, but the general arrangement is worth describing, since once it is understood the mechanism of replication is likewise apparent. The basic structure is shown in Figure 6. The molecule consists of two strands, each of which has a continuous sugar-phosphate-sugar-phosphate backbone, to which are attached the four bases. In any one strand the bases can occur in any sequence. Thus for example a strand 100 bases long could have any one of 4^{100} base sequences (there are 4 different bases, any one of which could occupy any one of the 100 sites). This enormous variety of base sequences makes it possible for DNA to meet the first requirement listed above; this is sometimes expressed by saying that the specificity of DNA resides in the sequence of bases.

But although a given strand can have any base sequence whatever, there is a fixed relationship between the sequences of two complementary strands. This arises because only two types of pairing are possible between a base on one strand and a base on the other; adenine always pairs with thymine and guanine always pairs with cytosine. This fact at once suggests how replication of a DNA molecule might take place (Figure 7). Suppose a length of DNA molecule separates into its two component strands. Then the bases will pick up their appropriate partners, which are already present in the cell. The result will be two DNA molecules, each identical in sequence to

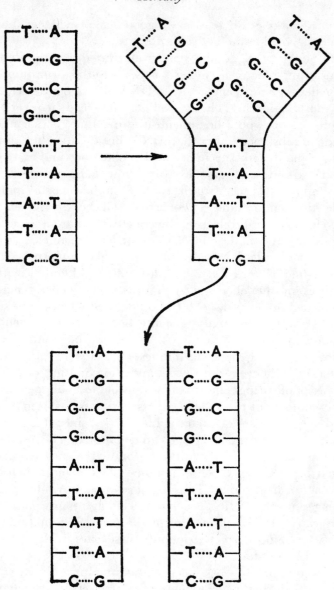

Figure 7. The replication of DNA.

the original one. There is now a good deal of evidence that this is how DNA does in fact replicate.

Although it is a big step from the discovery that the transforming factor in bacteria is DNA to the assumption that all genes are simply molecules of DNA, this step has been taken by most biologists. The reason for this ready conversion is interesting. It would be difficult to prove that genes cannot be made of substances other than DNA. But the structure of DNA does explain both specificity and replication, and, as will be described later, it also explains how genes can influence development. It seems unlikely that another class of molecule able to combine these properties will be discovered and, therefore, natural to leap to the conclusion that all genes in all organisms are made of DNA (or at least of nucleic acid; a related molecule, RNA, can have gene-like properties).

The line of work starting with Griffith and ending with the analysis of the chemical structure of the genetic material proceeded side by side with two other lines of work. The first of these – the genetic analysis of the fine structure of genes – is beyond the scope of this book. But it can be summarized as follows: purely genetic techniques – the crossing of organisms differing in one or more ways and the counting of different classes of offspring – has made it possible to analyse genes into their component parts. The greatest progress has been made in studying the genes of bacteria and viruses, since it is here that the largest number of offspring can be counted. The result of the analysis has been to show that not only are genes arranged in linear order along the chromosomes, but the genes themselves have a linear structure. Thus a gene can be shown to consist of a long sequence of 'sites', such that a change at any one of these sites can alter the properties of the gene. It is natural, and almost certainly correct, to identify these 'sites' with the bases of the DNA molecule.

A second line of work has been an attempt to discover what genes do during development. The answer to this question might seem to be: 'Genes do many different things; for example, they may affect the shape of one's nose or the colour of one's skin or whether one's blood will clot; there is no one thing that genes do.' This is true enough, yet genes all do have something in common; this has been summed up in the slogan 'one gene, one enzyme'. An enzyme is a protein molecule (what a protein is will be explained in a moment) which acts as a catalyst – that is to say, which enormously speeds up some particular chemical reaction which in the absence of the enzyme would proceed very slowly. The complex series of chemical reactions which go to make up the metabolism of living things depends on the presence of large numbers of different and specific enzymes, each catalysing a different step.

The word 'protein' must now be explained. Protein molecules, like those of DNA, are long strings, but in this case the strings are single and not double, and may (as in the so-called globular proteins) be folded up so as to form a blob. The string is formed of a series of smaller molecules, the amino acids, which can be thought of as resembling the poppets of a necklace. Each amino acid has a basic end (chemically, an NH_2 group) and an acid end (chemically, a $COOH$ group), and the basic end of any one amino acid can become attached to the acidic end of any other, with the elimination of a water molecule. Thus long strings of these amino acids can be built up; these strings are proteins. Many different kinds of amino acids exist, but only twenty occur in proteins. There appears to be no restriction on the order in which amino acids can be arranged. Consequently, as in the case of DNA molecules, an inconceivably vast number of different proteins are possible. The chemical specificity of an enzyme – that is, the chemical reactions it will catalyse and the conditions in which it will do so – depend on the sequence of its component amino acids.

We can now turn to the idea, 'one gene, one enzyme'. The evidence for the idea will be discussed in Chapter 5. For the present, it is sufficient to explain what is being suggested, which is as follows:

(i) The presence of any particular enzyme in an organism depends on the presence of a particular gene, and
(ii) what genes do, and the only thing they do, is to determine enzymes. If a gene has other effects as well, these effects are

secondary; if for example a gene influences the shape of the nose, it does so by determining the presence of a particular enzyme, which in turn influences the shape of the nose.

These suggestions can be accepted only after certain reservations have been made, but the general idea that what genes do is to determine the specificity of enzymes is true and important. The reservations are as follows:

(i) Not all proteins are enzymes. For example, some, like the proteins which form silk or hair, are structural. These non-enzymic proteins also need genes to specify them.

(ii) Some proteins (for example, haemoglobin – see pages 98–104) are formed by bringing together in the cytoplasm of the cell two or more components determined by different genes.

(iii) It is known that in bacteria some genes act only to 'regulate' the activities of other genes and the same is likely to be true in higher animals and plants. This subject will be discussed further in Chapter 5.

Despite these reservations, the general picture that genes determine enzymes is true. Since it is known that proteins consist of specific sequences of twenty kinds of amino acids, it is natural to suppose that genes act by determining the sequence in which amino acids are strung together to form proteins. This supposition is fairly certainly correct; the details of the process are described in Chapter 5.

To summarize the argument of this chapter, it has been suggested that:

(i) Many of the inherited differences between organisms are caused by Mendelian factors or genes. In sexually reproducing organisms, each parent transmits one complete set of constituent genes to the new individual.

(ii) Genes are small parts of chromosomes.

(iii) Genes are molecules of DNA, and owe their specificity to the order in which their constituent bases are arranged. The double structure of DNA, together with the restriction of base pairing to adenine-thymine and guanine-cytosine, provides a mechanism for the exact replication of DNA molecules.

(iv) The primary action of genes is to determine the presence of specific proteins.

CHAPTER 4

Weismann, Lamarck, and the Central Dogma

We are now in a position to discuss two difficulties which together form the central problem of evolutionary theory, at least in its genetic aspect. First, if like always begets like, there can be no novelty in evolution; and second, not all differences between individuals are due to their genetic make-up.

For example, Japanese are on the average shorter than Americans or Englishmen. However, children born to Japanese immigrants in the United States grow on the average to be taller than native Japanese. The difference in stature between Japanese and Americans is therefore due partly to environment, probably to nutrition. Similarly, the red deer of Scotland or the west of England are smaller than red deer kept in parks. When Scottish red deer were released in New Zealand, their descendants grew to be as large as deer from parkland.

It does not follow that there are no genetic differences in stature between men, or in size between red deer. In the case of human stature, for example, there is considerable resemblance between fathers and their sons, even when the sample is confined to a group living in fairly uniform social conditions, and there is a still closer resemblance in stature between monovular twins, i.e. a pair of twins who have been developed from a single fertilized egg.

In general, then, the observable characters of an organism, its phenotype, result from the interaction of environment and genetic potentiality. This is true even for many gross departures from the normal phenotype, such as the dumpy wings of *Drosophila*. It was stated earlier that flies with at least one *D* gene develop normal wings. This is true in the great majority of environments. However, if normal D/D individuals are exposed to a high temperature (40 °C.) for several hours at the correct stage of wing development in the pupa, a large proportion of them will develop dumpy wings. This should perhaps be regarded as a result of damage during development, of which many examples could be given. Nevertheless the examples of adaptive developmental flexibility discussed in Chapter 1 show that a zygote of a given genetic make-up has the property, not of always developing a particular phenotype whatever the conditions, but of developing a range of phenotypes according to varying conditions.

This idea is expressed in the definition of heredity by the Russian botanist Lysenko: 'Heredity is the property of an organism to require certain conditions for its life and development, and to respond in definite ways to various conditions.' This definition has the merit of emphasizing the relationship between organism and environment, and the flexibility of development. However, most geneticists would prefer it as a definition of 'nature' or 'constitution' rather than of heredity, which implies a resemblance between parent and offspring. Lysenko himself believes that if an organism is reared in changed conditions, and in consequence develops along a different path, then, at least in some cases, its offspring also may tend to develop along the new path.

This is the theory which has rather inexactly been called 'the inheritance of acquired characters'; it was the view held by Lamarck and accepted by Darwin. There is no theoretic reason why the environmental conditions of the parent should not affect the nature of the progeny, and there are a number of environmental stimuli, often of rather extreme character, which do have such an effect. In such cases, however, the kinds of change produced in the offspring do not resemble any changes produced in their parents due to the direct action of the environmental stimuli. It is in general true that quite striking changes can be produced in the phenotype of individuals by changed conditions, without affecting the nature of their progeny. For example, if two flies which have developed dumpy wings as a result of heat shock during their pupal life are crossed, their offspring develop normal wings, unless they too are exposed to a heat shock.

Such results show that the processes which lead to the production of gametes, and hence which contribute to the nature of the next generation, are to some extent independent of changes in the body, so that the hereditary properties of the gametes are not easily influenced by the environmental conditions in which the animal or plant is kept. The most extreme expression of this view was due to Weismann, who regarded the fertilized egg as the starting point of two independent processes. One leads by cellular division and differentiation to the individual body, or 'soma', which can be modified by external conditions, and which is mortal. A second process of cellular division in the 'germ line' gives rise to the sex cells, and hence to the next generation. Thus the germ line is potentially immortal; Weismann held that it was also independent of changes in the soma.

Now in many animals it is possible at a very early stage in development to distinguish between germ line and soma. Certain cells, known as primordial germ cells, can be recognized, mainly by their retention of the unspecialized appearance of early embryonic cells and their failure to develop the special features of for example, muscle, bone, or nerve cells. It can be shown that these primordial germ cells are incorporated in the ovary or testis, divide, and ultimately give rise to eggs or sperm. This early segregation of germ line and soma is not apparent in all animals, and does not occur in plants, in which any cell from the growing point of a shoot is capable of giving rise to sex cells.

The early separation of the germ line in many animals requires an explanation, but it does not by itself prove the germ line to be independent of changes in the soma. In order to grow and divide, cells in the germ line must be supplied with substances elaborated in other parts of the body. The same is true of chromosomes and genes. Thus in saying that the germ line is independent of the soma, Weismann could hardly have meant that there is no flow of energy or material from soma to germ line.

Weismann's meaning can today be expressed more precisely

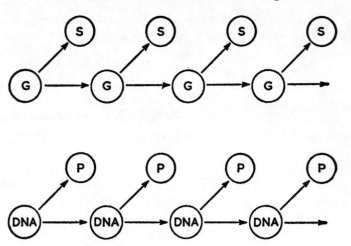

Figure 8. Weismann and the central dogma.

in molecular terms. In Figure 8a Weismann's theory is shown in diagrammatic form, and in Figure 8b is shown what Crick has called the 'central dogma' of molecular genetics, a dogma which states that information[1] can flow from nucleic acids to proteins, but cannot flow from protein to nucleic acid.

The connection between the two diagrams is clear. Accepting for the moment the truth of the central dogma, it could account for the correctness of Weismann's views in the following way. If an organism is raised in a new environment, this may alter the relative amounts or dispositions of different types of protein molecules, in such a way as to render the organism better able to survive the new conditions. But, if the central dogma is true, this cannot cause an equivalent change in the hereditary material or DNA, and so cannot cause the adaptation to be transmitted to the next generation.

---

[1] 'Information' is a technical term borrowed from communication engineering. Its use in biology is full of pitfalls. Its meaning in the present context can be made clear by the following rather pompous formulation of the central dogma: 'If a change is made in the sequence of bases in a nuclei acid molecule of a cell, this can cause the production of protein molecules with a changed sequence of amino acids, but if the sequence of amino acids in a protein is changed, this will not cause the production of a nucleic acid molecule with a new sequence of bases, itself capable of causing the production of additional protein molecules of the new kind.'

It follows that Weismann's views must be accepted and Lamarck's rejected, provided that two points can be established:

(i) that the central dogma of molecular biology is true, and
(ii) that changes in the structure of organisms induced by changes of their environment cannot be transmitted direct to the next generation, without first being 'translated' into nucleic acid.

These two points are respectively equivalent to showing that, in Figure 8, there is no arrow from P to DNA, and no arrow from P to P. At present, there is no reason to doubt the truth of the central dogma, but, as I shall explain below, there is at least one group of experiments suggesting a hereditary mechanism not dependent on nucleic acids.

The best reason for accepting the truth of the central dogma is as follows. The process whereby information is passed from DNA to protein is now well understood in chemical terms (see pages 92–3), and can be carried out in a test tube in the absence of living cells. Not only does the process not work backwards, it is difficult to see how it could conceivably do so. Of course this argument is not decisive, but my guess is that the central dogma is here to stay.

Before discussing whether there are hereditary mechanisms not dependent on nucleic acids, we have to answer the following question: if changes in the hereditary material are not the consequence of individual adaptation, what is the origin of evolutionary novelty?

If Figure 8 is accepted as an adequate model of heredity, the only way in which evolutionary novelty can arise is by changes in the structure of DNA molecules or by errors in the process of DNA replication; such events are called mutations. Mutations are known to occur spontaneously – i.e. without our doing anything deliberately to cause them – with low frequency. It was shown by Muller that their frequency is greatly increased by X-rays. Since that time, a number of chemical substances have been found which increase the frequency of mutation.

More important, different chemical and physical agents produce different types of change. There is nothing particularly surprising about this. For example, one class of mutagenic substances is the so-called 'base analogues'. These are molecules which bear a close chemical similarity to one of the four bases, adenine, thymine, guanine, or cytosine. When such analogues are present, a replicating DNA molecule may incorporate one of them instead of the corresponding base, the result being a mutation. Thus a particular analogue would be expected to cause mutations at particular sites within the gene, and this has been shown by Freese to be the case in viruses.

Thus it is no longer possible to think of mutations as 'random'. But we can abandon the concept of the randomness of mutation without accepting Lamarckism, and while continuing to hold that it is selection and not mutation which determines the direction of evolution. The important points are:

 (i) Most mutations lower the fitness of the organisms carrying them. If this were not so, mutation by itself without natural selection could account for evolution.
(ii) Mutations do not adapt organisms to the agent which produced them. For example, a bacterium carrying a mutation caused by X-rays will usually not be more resistant to X-rays; indeed, it is not more likely to be resistant to X-rays than if the mutation had been caused by some other agent.

We must now consider the possibility that, although the central dogma may be true, Weismannism is false because changes induced in the structure (protein or otherwise) of an organism can be replicated and transmitted to the next generation without involving nucleic acids; in other words, there may be a second genetic mechanism. Before discussing the detailed evidence on this point, it is worth indicating what has to be shown. It must be shown that in a population of organisms of type A, which beget organisms like themselves, there can arise an organism of type B, differing from A in some way not involving its nucleic acids, and that type B will beget others like itself. Thus merely to argue that organisms need cytoplasm (i.e. material other than the nucleus) in order to reproduce does not demonstrate the existence of a genetic mechanism in the cytoplasm. It must be shown that the two types of cytoplasm can maintain their own characteristics, although their nucleic acids are identical.

Since eggs contain a large amount of cytoplasm and very little sperm, this kind of inheritance is suspected when the characters of the offspring of a cross between different strains depend upon which way the cross is made, resembling in either case those of the female parent. Discussion of cytoplasmic inheritance is made more difficult by the fact that the term can be used to describe several quite different phenomena. Three examples will now be described, which have in common only that the properties of the egg cytoplasm influence the characters of the individual developing from that egg, but which are in other respects quite unlike one another; most geneticists would prefer to use the term 'cytoplasmic inheritance' only in situations resembling the third of these examples.

(a) *Delayed Gene Action.* Hens lay eggs of different colours, but the eggs laid by any one hen all have shells of the same colour. The colour of the egg shell is genetically determined, but it is determined by the genotype of the hen that lays the egg, and not of the chick contained within it. If we regard the colour of the shell as part of the mother's phenotype, then the genetic mechanism controlling it is just the same as that controlling, for example, the colour of her plumage. If however, we regard the colour of the shell as part of the phenotype of the chick, then the mechanism would have to be described differently, the phenotype of the chick being determined by the genotype of the mother. In this case, the former of the two ways of looking at the situation is the simpler one. There are, however, situations in which the characters of an adult individual are determined by the genotype of the mother, and not by its own genotype.

For example, most pond snails (*Limnea*) have shells coiling to the right (dextral), but occasional individuals are found with shells coiling to the left (sinistral). The eggs laid by a single snail develop into individuals which are all alike, but not necessarily resembling their mother. Thus the direction of coiling is determined by some property of the mother which influences the cytoplasm of the eggs which she produces, and thereby the coiling of her offspring. It has been shown that this property of the mother is itself determined by her own chromosomal genotype. Thus if individuals are classified, not as dextral and sinistral, but as 'producing dextral off-spring only' and 'producing sinistral offspring only', then this latter characteristic is inherited as a simple Mendelian one. The only difference between this situation and more typical examples of Mendelian inheritance is that in this case the genes in a zygote influence, not the development of the individual, but the properties of the eggs which it produces. Such cases are interesting, but do not show the existence of a non-nuclear hereditary mechanism.

(*b*) *The Transmission of Environmentally Induced Changes*. Locusts exist in two phases, solitary and gregarious. These phases differ both in behaviour and in structure; there is no sharp distinction between them, but all intermediates are found. In nature, it is the gregarious forms which, arising in areas where the population has reached a high density, migrate in vast swarms, and are responsible for locust plagues.

The development of a locust is greatly influenced by the conditions in which it is raised, and in particular by the presence or absence of other locusts. In crowded conditions, individuals tend to develop into the gregarious phase. However, development is also influenced by the environmental conditions of the mother. Eggs laid by solitary females tend to develop into the solitary phase, although they can be caused to develop at least some features of the gregarious phase if they are kept crowded. Thus it takes two or more generations to obtain fully gregarious individuals from solitary parents, or vice versa.

Here the conditions in which a female is raised influence not only her own phenotype, but also, via the cytoplasm of the eggs she lays, the phenotype of her offspring. The phenomenon is perhaps best thought of as a form of developmental flexibility which is in part inherited. But the stability of the cytoplasmic states is too low for them to be of evolutionary importance.

(c) *Cytoplasmic Inheritance.* The green colour of plants is due to the presence of the pigment chlorophyll, which is contained in structures called 'chloroplasts' visible in the cell cytoplasm. These chloroplasts have many of the properties of chromosomes. It is probable that new chloroplasts can arise only by the division of previously existing ones; in lower plants this division can be observed directly, but in higher plants propagation involves a small colourless phase. Chloroplasts are transmitted to the progeny by the female parent, and occasionally also in pollen. They differ most obviously from chromosomes in that they are present in large numbers in a single cell, whereas a particular chromosome is present only twice in a somatic cell and once in a gamete.

Some differences between chloroplasts can be shown to be due to nuclear genes. Thus in *Primula sinesis*, there is an albino variety, lacking chlorophyll, which shows typical Mendelian inheritance. The homozygous white form is inviable, but the heterozygote is yellow and can be used for breeding experiments. However, in the case of another yellow variety of *Primula*, inheritance is wholly maternal; plants resemble their female parent in colour, and are unaffected by their male parent. Here it is clear that the chloroplasts are not only self-reproducing, but also retain their special qualities in the course of reproduction, as do chromosomes.

Thus in the case of chloroplasts, the criteria for demonstrating a non-nuclear genetic mechanism have been met. The first two editions of this book continued at this point 'but it does not follow that the mechanism does not depend on nucleic acids; it is at least possible that it depends on the presence of nucleic acids in the cytoplasm.' It is now known that chloroplasts not only contain DNA, but also the machinery (see pages 92–3) for translating this DNA into protein. Of the chloroplast proteins, some are coded for by the chloroplast's own DNA and translated *in situ*, others are coded for by nuclear genes. The same is true of another organelle, the mitochondrion, which is present in animal as well as plant cells; the possible evolutionary significance of these facts will be discussed on pages 118–21.

There is, however, a series of experiments by Sonneborn on the slipper animalcule *Paramecium* which establishes both the inheritance of an acquired character and the existence of a hereditary mechanism not dependent on nucleic acids. The surface of *Paramecium* is covered by an intricate pattern of small hairs or 'cilia'. If the arrangement of these cilia is altered, either accidentally or by surgical interference, the alteration may be transmitted indefinitely to the descendants through many hundreds of cell divisions. Alterations transmitted in this way may be minor (a single row of cilia beating in the wrong direction) or major (the presence of two gullets instead of one). The transmission can be shown to be independent of changes in the nucleus. There has been much argument about whether there is DNA associated with the bases of the cilia. At present it looks as if there is not, although a related nucleic acid, RNA, is present. But in any case the answer is irrelevant. It is absurd to suppose that if one churns around in the cuticle of *Paramecium* with the point of a needle and in so doing alters the orientation of some DNA molecules, this alteration could be transmitted indefinitely by virtue of the precise complementary base pairing of replicating DNA. Sonneborn seems to have demonstrated the existence of some larger-scale structural arrangement in the cuticle of *Paramecium* with hereditary properties.

Thus we cannot rule out the possibility of genetic mechanisms not dependent on the replication of nucleic acids. Yet it seems unlikely that they have been of major importance in evolution. The vast majority of inherited differences between organisms which have been analysed have turned out to be caused by differences between nuclear genes. It also seems clear that mutations, although not random, do not adapt the organism to the agent which caused them, and that a gene, once mutated, replicates in its changed form with the same degree of accuracy as it did in its original form. It is this feature of the genetic system which has made some biologists reluctant to accept the Mendelian scheme. Most organisms can and do change adaptively in response to changes in the environment, and most organisms appear to resist inadaptive changes, and to revert to their former state if the distorting influence is removed. Organisms can adapt and are homeostatic, whereas genes cannot adapt and are not homeostatic.

Yet once it is realized that the genetic system is itself the product of evolution by natural selection, its peculiar properties make sense. Most 'acquired characters' are the consequence of injury, starvation, disease, or senescence – only a minority are adaptive. The inheritance of acquired characters would lead to deterioration rather than evolution. On the other hand homeostatic genes, resembling the self-correcting codes used for programming computers, which always reverted to their original form after mutating, would make evolution impossible.

CHAPTER 5

Molecular Evolution

(i) *Genes and Proteins*

If the account of heredity given earlier is correct, it follows that the whole pageant of evolution since pre-Cambrian times – ammonites, dinosaurs, pterodactyls, mammoths, and man himself – is merely a reflection of changed sequences of bases in nucleic acid molecules. What is transmitted from one generation to another is not the form and substance of a pterodactyl or a mammoth, but primarily the capacity to synthesize particular proteins. The development of specific form is a consequence of this capacity, and the capacity itself depends on the self-replicating properties of DNA.

Even if this account is not the whole truth, it is clearly a very important part of the truth. In this chapter I shall discuss some of the implications; but first, I must explain why it is thought that the primary function of genes is to specify proteins.

The idea 'one gene, one enzyme' originated from a study of the eye pigments of *Drosophila*. Normal eyes are dark red, containing both red and brown pigments. Many mutants are known which cannot form the brown pigment, and which therefore have light red eyes. Beadle and Ephrussi developed a new technique for analysing these mutants. They found that if they removed from a larva the small ball of cells destined to become the adult eye, and grafted it into the body cavity of another larva, then when the 'host' larva metamorphosed, the rudiment turned into an adult eye. Then, for a number of bright red mutants, they grafted mutant eye rudiments into the body cavity of genetically normal larvae. For most mutants, the

87

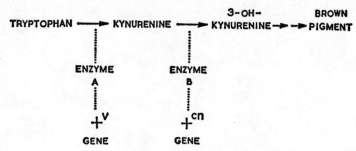

Figure 9. Eye colour genes in *Drosophila*. The chemical formulae of the substances at the top are unimportant; what matters is that each arrow represents a simple chemical transformation brought about by a single enzyme.

behaviour of the graft was 'autonomous'; that is to say, it developed the same bright red colour as it would have done in its original site. But two mutants, *v*, 'vermilion', and *cn*, 'cinnabar', were non-autonomous—*v* and *cn* rudiments in a wild-type host developed dark red eyes. This suggested that *v* tissues are unable to make some substance which can be made in normal tissue, and that this substance could diffuse from the normal tissue into the graft; *cn* tissues are unable to make a different diffusable substance.

Clearly these substances cannot themselves be genes, since genes could hardly diffuse; they must be products of gene action. Their identification was difficult, but the conclusions finally reached are shown in Figure 9. Tryptophan is a necessary pre-cursor of brown pigment. It is changed first into kynurenine and then into hydroxy-kynurenine, and then, after a number of further steps, into brown pigment. The first of these transformations depends on the presence of the wild-type allele of the vermilion gene, $+^v$, and the second on the wild-type allele of the cinnabar gene, $+^{cn}$.

It was already known that chemical steps of this kind in living systems depend on the presence of specific enzymes. This might have suggested that genes are in fact enzymes. One difficulty with such a view is that enzymes are known to be active in the cytoplasm of cells, outside the nucleus. When later it became clear that the genetic material was DNA, it was supposed that specific DNA molecules determined the presence of specific enzymes.

These experiments on two *Drosophila* mutants were clearly insufficient by themselves to establish the principle, 'one gene, one enzyme'. This became generally accepted as a result of the work of Beadle and Tatum on the fungus *Neurospora*. In this fungus – and subsequently in many other microorganisms – it has proved possible to isolate genetic mutants which differ from the normal only in their inability to perform a single chemical step – i.e. in the absence of a single enzyme. It is roughly true to say that whenever biochemists have been able to identify an enzyme, it has been possible to find mutant strains lacking just that enzyme; indeed many new enzymes have been discovered by the study of genetic mutations.

The implication is that an unmutated wild-type gene has as its function the determination of a specific enzyme, and that all enzymes require a gene to specify them. It does not of course follow that all genes determine enzymes; exceptions to this generalization are listed on page 74. But most biologists today think that all genes either act to specify a particular protein, or play some other role in the control of protein synthesis. If so, all other genetically determined differences – in the shapes of our noses or in the degree of our intelligence – result from differences in the capacity to synthesize proteins.

(ii) *The Coding Problem*

How are genes able to specify proteins? As explained earlier, a gene is a molecule of DNA consisting of two complementary strings of bases, and a protein consists of a string of amino acids. The problem of how the one determines the other is often called the coding problem. It can be split into three subsidiary questions, as follows:

(A) What *kind* of code is it?
(B) What is the machinery of translation?
(C) What is the dictionary?

A. *What kind of code is it?* This question is best explained by an example. Let us suppose we want to code a message written in the roman alphabet of twenty-six letters into the ten digits 0 to 9. One obvious way would be to represent each letter by a pair of digits – for example A = 01, B = 02 ... Z = 26. If we wanted to make it slightly more difficult to crack, we could allow several pairs of digits to stand for a single letter—for example 17, 73, and 91 could all be read as A, and similarly for the other letters. Such a code would be 'degenerate'; knowing the code, it would be possible to translate a message in digits unambiguously into the alphabetic version, but not vice versa.

A doublet code of this kind is not the only possible one. For example, there are 676 different pairs of letters, and 1000 triplets of digits, so we should get a more compact code by using a different triplet of digits to represent each doublet of letters. There are many other possibilities.

Another difficulty is knowing where to start transcription. Suppose for example we had a simple non-degenerate code with A = 01, ... Z = 26. How should we translate a section of code 1201–10216 ... ? Is it 1.20.11.02.16 ... (i.e. **TKBP**) or is it 12.01.10.21.6 ... (i.e. LAJU)? There are two possible ways of answering this question. One is to notice that no meaningful doublet can start with a 6; therefore the second version is wrong, and the translation is **TKBP**. An alternative would be to know where the message started and to count off in twos. A practical difficulty arises with the second method if the message is a very long one (and the genetic message is very long indeed) – a single mistake in counting off would make the whole of the rest of the translation wrong. This difficulty could be largely overcome by dividing the message into 'words' separated by 'stops'; for example, 99 could be used to symbolize the end of one word and the start of another.

Thus in asking 'what kind of code is it?' we are asking the following kinds of questions:

Is each amino acid represented by a separate group of bases, and if so how many?

Is the code degenerate – i.e. do several different groups of bases represent the same amino acid?

How does the cell 'know' where to start transcription? Is this knowledge based on the recognition of 'nonsense' syllables, or are there 'stops' indicating where transcription should start?

But before tackling these questions, there is a difficulty to be cleared up. A DNA molecule consists of two complementary strands. Are both strands transcribed, or is the message carried by one only, the other being required for replication? It seems clear that, whatever the code, only one of the two strands can carry a message. Thus suppose that in the roman alphabet the letters were paired off, say *a* with *b*, *c* with *d*, *e* with *f*, and so on; this corresponds with the pairing of adenine with thymine and guanine with cytosine in DNA. Then a message – say 'England expects...' would have a unique complement – fmhkbmc fwofdst. We could recover the original message by forming the complement of this complement, bringing us back to 'England expects...' – this in fact is how DNA is replicated. But obviously we can only arrange for one of the two complements to make sense; the other, necessary for replication, will be nonsense. Thus there are strong theoretical reasons for supposing that the message would be carried by one of the two strands only. There is now some direct experimental evidence that this is true, but we have no idea what decides which strand shall be read.

There were also strong *a priori* reasons for expecting each amino acid to be represented by a group of three bases. The simplest kind of code is one in which each symbol in one alphabet is represented by a fixed number of symbols in the other. The smallest number of bases required to code 20 amino acids is three. Since there are 4 bases, and hence only 16 different doublets, a doublet code would therefore leave 4 amino acids uncoded. But a triplet code, with 64 different triplets, is ample to code 20 amino acids.

The answers to the questions on the previous page are now known; they are as follows:

Each amino acid is represented by a group of three bases, or 'triplet'.

The code is indeed degenerate. Some amino acids are represented by one triplet, others by as many as six.

There are 'stop' triplets. The cell 'knows' where to start transcription, and counts off in threes.

The methods used to reach these conclusions would take too

long to explain. They depended on the existence of mutations which either inserted or deleted a single base in a gene. Clearly, if the cell does indeed count off in threes, such mutations (called, for obvious reasons, 'frame shift' mutations) would cause the cell to mistranslate everything beyond the mutation.

B. *What is the mechanism of translation?* A code usually implies the presence of a decoder, but the cell has to do its own decoding. The cell's decoding machinery is complicated, but it must be described, even if in a rather dogmatic way, because a discussion of how the code may have originated or of how it may have changed during evolution depends on an understanding of this mechanism. More important, the 'central dogma' depends on the irreversibility of this machinery. In describing the process three new classes of molecule will be mentioned:

   (i) 'Messenger RNA', whose function it is to carry the message from the nucleus to the cytoplasm, and there act as a template for the synthesis of a protein molecule.
 (ii) 'Transfer RNA' molecules, and
(iii) 'Activating enzymes', which between them ensure that the appropriate amino acids are lined up opposite the corresponding triplets.

RNA, or ribose nucleic acid, is a molecule similar to DNA, but differing in two respects; first, it is single-stranded, and second, one of the bases of DNA, thymine, is replaced in RNA by the base uracil. A 'hybrid molecule' can be formed by one strand of a DNA molecule pairing base by base with an RNA molecule. Thus just as in replication each strand of the DNA molecule can form a complementary strand against itself, so when a gene (DNA molecule) is active, the two strands separate momentarily and the one strand which carries the message forms a complementary RNA strand against itself. This strand, the messenger or mRNA molecule, then moves into the cytoplasm, to one of the bodies called ribosomes which are the sites of protein synthesis

It is now necessary that the appropriate amino acid – let us say proline – be brought to the appropriate triplet in the mRNA molecule, which for proline is thought to be a triplet of three cytosine bases. This is done in two stages. First the proline molecule is attached to a particular 'transfer RNA' molecule, and then this transfer molecule attaches itself to the appropriate site on the messenger. The latter step probably takes place because this particular transfer molecule has an exposed triplet of guanine bases which will pair with the three cytosine bases of the mRNA. But the attachment of the proline molecule to this particular transfer molecule – and not for example to a transfer molecule with an exposed triplet of adenines – is a more complicated job, and requires the existence of a particular activating enzyme. Thus for each kind of coded triplet, there exists a special transfer RNA molecule which will pair just with that triplet, and also a special activating enzyme which will attach one kind of amino acid, and only one kind, to that transfer molecule. It is the function of the transfer molecules and activating enzymes to line up the amino acids opposite their appropriate triplets in the messenger molecule. The amino acids are then joined up end to end to form the completed protein.

Molecular biologists have given some useful and expressive names to various aspects of this process. The first stage, the formation of messenger RNA molecules complementary to one strand of the DNA, is called 'transcription'; the second stage, occurring on the ribosome, whereby the appropriate amino acids are lined up opposite the corresponding triplet, is called 'translation'. The triplet of bases in a messenger specifying an amino acid is a 'codon', and the triplet of bases in a transfer molecule which pairs with the codon is an 'anticodom'. Less officially, the two strands of the DNA molecule are 'Watson' and 'Crick', but as yet there is no international agreement as to which gets transcribed and which is there only to help in replication.

C. *What is the dictionary?* Which triplets code which amino acids? Also – more interesting to an evolutionist – is the code the same for all organisms?

The answers to these questions depend on the discovery by Nirenberg that protein synthesis occurs in cell-free extracts in a test tube, and that such extracts can be 'primed' with messenger molecules of known base composition; it is then possible to discover which amino acids have been incorporated into proteins. Thus if an extract is primed with mRNA consisting of a string of uracil bases only, the protein recovered consists of a string of phenylalanine molecules. It follows that the triplet UUU in mRNA codes for phenylalanine. By an elaboration of this method it has been possible to assign a meaning to every codon, as shown in Figure 10.

The first important conclusion which has been drawn is that the code is universal; that is, the dictionary is the same for all organisms. One reason for believing this is as follows. The cell-free extracts described above contain ribosomes, transfer RNA molecules and activating enzymes. If primed with synthetic mRNA of a particular base composition, they synthesize proteins containing the same amino acids, regardless of whether the extracts were derived from bacterial or mammalian cells.

This universality of the code does *not* arise because no mutation altering it is possible. Thus the correct assignments depend on specific activating enzymes capable of attaching the 'right' amino acid to the 'right' transfer molecule. These activating enzymes are themselves specified by genes; a mutation in such a gene can produce an enzyme which attaches the 'wrong' amino acid to a transfer molecule. Hence a mutation can alter the code. Such mutations have been identified and extensively studied in bacteria. The code is universal because changes in it are selected against. Thus it might be an advantage to replace leucine by valine in one particular site in one particular protein, but to alter leucine to valine in thousands of different proteins simultaneously would certainly be disastrous. It follows that the code, once evolved, cannot easily change. In fact, a codon could only change its meaning in an organism which first evolved so as never to use that codon in its genetic message; this seems not to have happened.

A further question to be asked is, to what extent is it

SECOND BASE

		U		C		A		G		
		UUU	Phe	UCU		UAU	Tyr	UGU	Cys	U
	U	UUC		UCC	Ser	UAC		UGC		C
		UUA	Leu	UCA		UAA	OCHRE	UGA	OFAL	A
		UUG		UCG		UAG	AMBER	UGG	Tryp	G
		CUU		CCU		CAU	His	CGU		U
	C	CUC	Leu	CCC	Pro	CAC		CGC	Arg	C
		CUA		CCA		CAA	GluN	CGA		A
		CUG		CCG		CAG		CGG		G
FIRST BASE		AUU		ACU		AAU	AspN	AGU	Ser	U
	A	AUC	Ileu	ACC	Thr	AAC		AGC		C
		AUA		ACA		AAA	Lys	AGA	Arg	A
		AUG	Met	ACG		AAG		AGG		G
		GUU		GCU		GAU	Asp	GGU		U
	G	GUC	Val	GCC	Ala	GAC		GGC	Gly	C
		GUA		GCA		GAA	Glu	GGA		A
		GUG		GCG		GAG		GGG		G

THIRD BASE

Figure 10. The genetic code. U, C, A, and G represent the four kinds of bases in RNA; the code is given in the form appropriate to messenger RNA. Phe, Leu, etc., stand for the amino acids phenylalanine, leucine, etc. The UAA, UAG, and UGA triplets operate as 'stops', marking the end of a gene.

arbitrary? Is there a good chemical reason why UUU codes for phenylalanine, or is it a matter of chance? There are in fact two different questions here. First, is the grouping together of a set of codons with the same meaning arbitrary? Clearly it is not. For example, all four codons with CU in the first two positions code for leucine. This feature of the code has the consequence of reducing the harmful effects of mutation. Thus a mutation from CUA to CUG would leave the protein unaltered. But it does not follow that this feature of the code evolved because of its effects on mutation; there may have been physico-chemical or historical reasons why sets of codons have acquired the same meaning.

Second, granted that codons are grouped in particular ways, is there any reason why a particular group of codons is assigned to a particular amino acid, or could CUA and CUG equally well have come to code for valine? We do not know. Certainly with the existing translating machinery there is no physicochemical reason why assignments should not be altered; conservatism depends on selection, not on chemistry. But we do not know how the code evolved in the first place, so that, although no one has suggested any convincing chemical reasons why particular codons have particular meanings, it would be rash to assume that no such reasons exist.

One final feature of the code is worth mentioning. If one measures the relative amounts of different amino acids occurring in. proteins, there is a very striking correlation between the amount of an amino acid present and the number of triplets coding for it. Thus leucine (6 codons) is approximately six times as abundant as tryptophan or methionine (one codon each). The correlation is not perfect (for example, there is too little arginine in protein), but is far too good to be an accident. Some possible explanations for the correlation are discussed on pages 105–6.

(iii) *The Evolution of Molecules*

The simplest genetic change is the alteration in the base sequence of a DNA molecule, resulting in a changed amino acid sequence of a protein. Enough is now known about the structure of a few proteins for us to see how this works out in practice. But before describing the results, a little more must be said about the structure of proteins. A protein is a string of amino acids arranged end to end; this arrangement is called the primary structure of the protein. If that were all, proteins would be long and string-like, whereas most enzymes are globular in shape due to the folding of the string. The string is first partially coiled into a long spiral – the so-called α-helix – to give a secondary structure consisting of a long cylinder. This cylinder is then itself usually folded up to give a globular tertiary structure. The secondary and tertiary folding, and hence the final shape, are themselves determined by the primary amino-acid sequence.

We can now consider the various possible changes which can

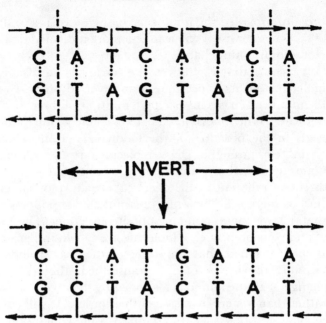

Figure 11. Consequences of a molecular inversion.

occur in DNA sequence, and their consequences for amino acid sequence:

(i)   Replacement of one base by another. Such 'point' mutations are the simplest mutational events; the usual result will be to replace one amino acid by another.

(ii) Deletion of a group of bases. If the number of bases deleted is a multiple of three, this will lead to the deletion of one or more amino acids from the protein (perhaps associated with the appearance of a new amino acid, if the beginning and end of the deletion do not correspond with the beginning and end of a triplet). But if the number of bases deleted is not a multiple of three, it will result in a change in all the amino acids following the deletion; radical changes of this kind are very unlikely to be advantageous.

(iii) Addition of a group of bases, probably by 'duplication' or copying of the same group of bases twice. As before, this will lead to the addition of a group of amino acids to the protein, provided the number of bases added is a multiple of three; otherwise it will lead to a more radical and almost certainly harmful change.

(iv) 'Inversion', that is, the removal of a section of a DNA molecule and its reinsertion in an inverted order. Even if the number of bases inverted is a multiple of three, this will not lead to the inversion of a comparable group of amino acids in the protein, for the following reason. The two chains composing a DNA molecule are 'polarized' in opposite directions; this simply means that there is a direction inherent in the chemical structure of the chain (just as there is for example in a necklace of poppets) and that the two chains point in opposite directions. The results of an inversion in such a molecule are shown in Figure 11. If it is assumed that only one of the two strands, for example the upper one, is transcribed, the result is a change of all the amino acids coded by the inverted region rather than a mere inversion of their order. As will emerge on page 130 this is a rather embarrassing conclusion.

No evidence exists, or in our present state of knowledge could exist, for the occurrence of short inversions. But that the other types of change have occurred in evolution has been demonstrated by a study of the haemoglobin and myoglobin molecules.

Haemoglobin is a protein found in the red blood cells of vertebrates, and able to combine reversibly with oxygen, and so help to transport oxygen from the lungs to the tissues. Myoglobin is a similar protein found in the muscles, where it acts to store oxygen and release it when needed. The myoglobin molecule is formed by the folding up of a single chain of amino acids. In adults, each haemoglobin molecule is formed from four chains, two of one kind, referred to as α chains, and two of another, β chains. Each of these chains folds up, and then the four molecules come to fit together like pieces of a three-dimensional jigsaw to form a molecule of haemoglobin, In foetal human beings, before birth, there is a different type of haemoglobin, which is likewise formed of four chains, two of which are α chains similar to those in adults, the other two being of a different type, the γ chain. There are good chemical reasons why a molecule formed of four components is a better oxygen carrier than a single molecule like myoglobin. There are also good reasons why there should be a different haemoglobin in foetus and adult; foetal haemoglobin must have a greater affinity for oxygen than adult haemoglobin if oxygen is to be transferred across the placenta from mother to child.

Thus four different kinds of amino acid chains are involved in the formation of the myoglobin and haemoglobin molecules – namely the myoglobin chain and the α, β, and γ chains of haemoglobin. Each chain is determined by a different gene; the four genes are thought to be on at least three different chromosomes. How have these four kinds of chains, and the genes which determine them, evolved? They have a history of some hundreds of millions of years, since fish, both bony and cartilaginous, have a myoglobin and a haemoglobin formed of four components. They are also among the few proteins for which complete amino acid sequences have been worked out. From these sequences it is possible to make plausible guesses at the evolutionary steps involved; it seems that point mutations, the addition or deletion of small groups of bases, and the duplication of whole genes, have all been involved.

The evidence for point mutations comes from a study of abnormal human haemoglobins. For example, people suffering from sickle cell anaemia (see page 175) have an abnormal adult haemoglobin, which differs from the normal only in that in the β chain one amino acid, glutamic acid, is replaced by another, valine, at a particular place. Other abnormal haemoglobins are known which differ from the normal only by the substitution of a single amino acid.

If the β and γ chains are compared, they are found to be the same length (146 amino acids) and to have 2/3 of their amino acids identical. The α chain is slightly shorter (141 amino acids); if its amino acids are paired off with those of the β chain, leaving gaps where appropriate, 40 of the pairs are identical. The resemblance between the α chain and myoglobin is less close but still very striking. It is just conceivable that the four genes responsible have each evolved from a wholly different ancestral gene by a process of convergent evolution. But it seems far more likely that three times in the course of evolution there has been a gene duplication (giving rise to a single chromosome with two identical copies of a gene which was present only once on an ancestral chromosome), followed by divergent evolution of the two copies, and by the separation of the two copies into different chromosomes. Processes which can give rise to duplication and separation are described on page 128.

The divergence of the genes determining the β and γ chains could have occurred by point mutation alone. But the divergence between the α and β genes requires the addition or deletion of small groups of bases, since the chains are of different lengths.

A highly simplified phylogenetic tree of haemoglobin chains is shown in Figure 12. It is assumed that the primitive jawless vertebrates (the Agnatha) had a single-chain globin molecule in their blood, because the surviving remnants of this group, the lampreys, have a single-chain globin. It is then supposed that at some time during the origin of the jawed vertebrates, approximately 450 million years ago, the globin-specifying gene duplicated, and the two copies diverged to become the genes for the α and β chains.

Since the amino acid sequences have been determined for a number of existing vertebrates, it is possible to estimate rates of evolution. First, by comparing the sequences of two chains, say carp α and mouse α, one can calculate the number of amino acid substitutions needed to convert one into the other. By dividing this number by twice the time to a common ancestor, in this case 900 million years, one obtains an estimate of evolutionary rate in terms of amino acid substitutions per polypeptide chain per year. Dividing again by the number of amino acid sites in the chain gives estimates of the number of substitutions per site per year. Some estimates of this kind are given in the table. It will be seen that they are surprisingly uniform; haemoglobin seems to have evolved at a rate of approximately one substitution per site per 10^9 years. Similar calculations can be made for a few other kinds of proteins, and the conclusion has been drawn that for a given type of protein

Average Rate of Evolution of Haemoglobin

Comparison	Substitutions per site per year $\times 10^{10}$
Human β *vs.* lamprey Globin	12·8
Human β *vs.* Human α	8·9
Human β *vs.* Other Mammal βs	11·9
Mouse β *vs.* Other Mammal βs	14·0
Human α *vs.* Carp α	8·9
Human α *vs.* Other Mammal αs	8·8

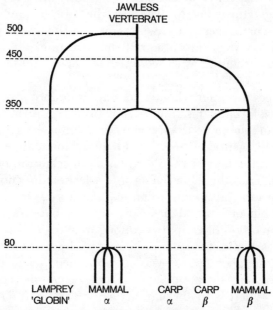

Figure 12. Evolutionary tree of the genes determining haemoglobin. The figures on the left are approximate times in millions of years.

the rate of evolution is approximately constant, but that rates are higher for some types of proteins than for others.

This conclusion is neither as clear nor as universally accepted as I have made it seem. There are uncertainties about the dates of common ancestors. Not all the estimates in the table are independent of one another; for example, the first two estimates in the table have a long period in common, but both are fully independent of the last estimate in the table. More serious, if the same amino acid substitution took place in two parallel lineages, the method described would ignore both occurrences, and so under-estimate the rate of evolution. An attempt has been made to correct for this effect in the estimates given in the table, but unfortunately the correction itself depends on assumptions about evolution which may not be true. Further difficulties arise from the redundant nature of the code. For these and other reasons, the conclusion that there is a uniform rate of evolution for a given class of protein is still being hotly debated. There is also disagreement as to whether the rate is uniform per generation or, as seems at present more consistent with the data, per year. My own impression is that the rate, although not absolutely constant, has varied much less than might have been expected.

Why does it matter whether the rate is uniform or variable? One reason is that if it is uniform then protein sequence data are a very valuable guide to phylogenetic relationships and to the time of divergence of lineages. Of more immediate interest, however, is the light it may shed on the mechanism of protein evolution. Thus the theory of natural selection does not predict a uniform rate (although it can explain it away); indeed, for haemoglobin one would have predicted a burst of evolution during the origin of viviparity among mammals; this does not show up as an acceleration in the rate of change of the α and β chains, although viviparity was accomplished by a further gene duplication, and the appearance of a new chain present only in the foetus. There is an alternative theory, the 'neutral mutation theory', which, with some additional but plausible assumptions, does predict a uniform rate. It was the proof of this prediction by Kimura, and independently by King and Jukes, which marked the birth of the theory. The argument depends on a rather subtle use of the concept of probability. Although subtle, the argument is both simple and elegant – some would say dangerously simple and elegant.

To fix ideas, consider the gene for the haemoglobin β chain. If each amino acid could change into any of the other nineteen, the total number of changes which could be produced by a single substitution in a chain of 146 amino acids is $19 \times 146 = 2,776$. However, because of the nature of the code, not all such substitutions could be produced by a single mutation substituting one base for another; for example, if you look at Figure 10 you will see that it needs at least two base substitutions to change phenylalanine to threonine. Single base substitutions can produce about 1,500 different changes in the protein. More drastic changes could be produced by the other types of mutation considered on page 97. Of these mutations, most would probably interfere with the functioning of the protein, and so would be eliminated by selection if they occurred. One or two might actually be favoured by selection. The neutral mutation theory proposes that some number, say R, of the possible mutations are selectively neutral; that is, they make so little difference to the protein that changes in gene frequency caused by natural selection are negligible compared to the accidental changes which are bound to occur in a finite population. The theory further proposes that most of the amino acid substitutions which occur in the evolution of proteins are neutral in this sense. The process has been called, for obvious reasons, non-Darwinian evolution.

I shall show in a moment how the theory can predict a uniform rate of evolution. But first I must emphasize some things the theory does *not* say. It does not say that most mutations are neutral, only that most of the mutations which are actually incorporated in evolution are neutral. Thus, of the 1,500 or so possible single amino acid changes in the β chain, it might be at some moment in evolution that 1,450 were harmful and only 50 neutral; all could occur in individuals, but only the 50 neutral ones would have any significant chance of being established in the population. Second, the theory does not say that all evolutionary changes are neutral. A minority of substitutions are of selectively favoured mutations, and it is this minority which is responsible for the evolution of adaptation. But, in addition to the Darwinian evolution of adaptations by natural selection, the theory proposes a second adaptively irrelevant type of change.

I now turn to the rate of evolution. Suppose there is a finite population of N individuals. Consider one haemoglobin β gene in one individual at the moment of conception; let the probability that a new selectively neutral mutation has arisen in that gene during the last generation be U; this is the 'neutral mutation rate'. Then the number of new neutral mutations occurring in the population per generation is $2NU$, assuming two genes per individual. Now what is the chance that any particular neutral mutation will ultimately be established as the only kind of gene in the population? The answer is $1/2N$, for the following reason. Of the $2N$ genes in the population at any instant, one is ultimately going to 'win', in the sense that all the genes in the population at some future time will be copies of one of the genes now. This is true in a finite population even if there is no selection, although it will take a long time if N is large. Now since the new mutation is neutral, if follows that all the $2N$ genes have equal chances of being the ultimate winner, so our particular gene has a $1/2N$ chance of being established, as stated above. Hence if there are $2NU$ new neutral mutations per generation, and if each has a $1/2N$ chance of being established, the rate of evolution is $2NU/2N = U$ substitutions per generation.

I have shown that the expected rate of 'neutral' evolution is equal to U, the neutral mutation rate. Why should U be constant? We know that $U = uR$, where R is the number of different kinds of neutral mutation possible, and u is the rate at which one particular mutation occurs. There is no reason why u should vary from species to species during evolution (there is also some argument about whether we would expect u to be constant per year or per generation). Hence the theory predicts a uniform rate of evolution provided that R remains constant.

The argument then proceeds as follows. In any given class of protein, there will be some regions which cannot change at all without serious consequences; in haemoglobin such a region would be that surrounding the haem molecule which combines reversibly with oxygen. There will be other regions where limited changes can occur without selectively significant effects. The number of such regions, and hence R, will be fairly constant for a given class of protein, but will vary from one class of protein to another.

Clearly this is a speculative argument. How well does it agree with what we know about protein evolution? Apart from the tendency for evolutionary rates to be constant, there is one other general conclusion which seems to emerge from sequence data: those proteins or parts of proteins which have some very specific function to perform (e.g. the active sites of enzymes, surfaces which have to 'fit' against other molecules) change little or not at all, whereas those parts which are subject to no very obvious selective constraints evolve rapidly. A good example of the latter is 'fibrinopeptide', a short region which is cleaved off from the protein fibrinogen so that the remainder of the protein can help clot the blood. This generalization is exactly what one would expect if the neutral mutation theory were true. However a strict selectionist would interpret the observations by saying that, for the kinds of molecules we have been discussing, natural selection has maintained the basic features of the molecule, while indulging in a kind of 'fine tuning' by altering those parts which can be altered without destroying the basic function. The implication would then be that the fine tuning is selectively important and not neutral.

At present there does not seem to be adequate evidence to enable one to come down firmly on one side or the other. One point is, however, worth making – the argument is a real one. It is sometimes suggested that both sides are right, since certainly some substitutions are selective, and some are neutral. This seems to me to miss the point, which is whether the proportion of neutral substitutions is large enough (well above 50 per cent) to account for the observed regularities in evolutionary rate.

I want, finally, to discuss a feature of molecular evolution which was first adduced as evidence for the neutral theory, but which, as was first pointed out by King, can also be explained in strictly selectionist terms. This is the close correlation between the number of codons for a particular amino acid, and the frequency of that amino acid in protein (see page 96). Our first problem is as follows: did the code evolve first and the amino acid frequencies in protein then change to agree with it, or was the evolution of the code influenced by the fact that some amino acids are more often useful in proteins than others? I cannot see how the latter could be true. This may merely reflect the fact that we do not in any case understand how the code evolved, but for the time being it seems reasonable to ask, given a fixed code, how do amino acid frequencies come to correspond?

Consider first the extreme selectionist position, which would be that, at any time in any species, either the existing protein is optimal (i.e. could not be improved by mutation) or there is one unique way of improving it. If this were so the frequencies of amino acids would be independent of the code, and would depend only on selective requirements. Now suppose that it is usually the case that, when a protein could be improved, there is more than one way of improving it, these ways being mutually exclusive. Then the substitution which actually occurs will be the one first arising by mutation. Now mutations towards leucine or arginine, each coded for by six codons, will be approximately six times as frequent as mutations to tryptophan or methionine, each coded for by one codon. Hence the frequencies of amino acids actually present will come to correspond approximately to the code. The observed correlation can thus be explained by a selective theory; it is not difficult to see that it could also be explained on the neutral theory. Hence the correlation does not enable us to decide between the two. However, it is worth remembering that *if* we accept the selectionist view that most substitutions are selective, we cannot at the same time assume that there is a unique deterministic course for evolution. Instead, we must assume that there are alternative ways in which a protein can evolve, the actual path taken depending on chance events. This seems to be the minimum concession the selectionists will have to make to the neutralists; they may have to concede much more.

It would be misleading to end a chapter on molecular evolution with a discussion of selectively neutral changes. The one point on which the two sides in the neutral versus selective controversy agree is that *some* changes are produced by selection, and it is these which are responsible for the evolution of adaptation. It is therefore important to ask whether genes can evolve so as to produce enzymes with new functions, and if so how. The work of Patricia Clarke and her colleagues on a bacterium, *Pseudomonas aeruginosa*, is beginning to answer this question. It has turned out that by applying appropriate selection pressures and by artificially increasing the mutation rate, it is possible to produce in the laboratory enzymes with new specificities.

Success has depended on an understanding of the mechanisms which control gene action. These mechanisms are described in greater detail on pages 122–4. For the present, the following points are sufficient. Most of the time, the gene for producing the enzyme which breaks down a particular substrate is kept switched off, so that no enzyme is produced. If the particular substrate is present, however, the gene is switched on and the enzyme is produced. This process of switching on and off has an obvious purpose in preventing the cell from making a lot of enzymes which are of no use to it. Now suppose that a strain of bacteria possesses an enzyme which can make use of a substrate S with high efficiency, and which can utilize a related substrate S^1 with very low efficiency. One might hope that if one kept such a strain on S^1 alone, and at the same time artificially increased the mutation rate, one might obtain a strain which could utilize S^1 with higher efficiency. Unfortunately, what usually happens in such experiments is that a mutation occurs in the regulator genes rather than in the genes producing the enzyme. A so-called 'constitutive' strain is obtained in which the enzyme-producing gene is permanently switched on. Enormous quantities of the original inefficient enzyme are produced, up to 20 per cent of the total protein of the bacterium. The population has adapted by producing more of an inefficient enzyme, not by evolving a better one.

Clarke studied an enzyme which breaks down a compound known as acetamide, and related compounds. Acetamide has a central chain of two carbon atoms; the related compounds have 3, 4, 5 and so on. Acetamide can be used as a source both of carbon and of nitrogen. Normal strains of *P. aeruginosa* can use acetamide and the 3-C compound; their enzyme can break down the 4-C compound with low efficiency (2 per cent), but cannot live on it because it does not switch the gene on; their enzyme cannot break down the 5-C compound at all. Starting with a 'constitutive' strain in which the enzyme-producing gene was switched on permanently, Clarke first obtained a mutant enzyme which could use the 4-C compound with 30 per cent efficiency. A further mutational step gave rise to a strain able to live on the 5-C compound, which could not be broken down at all by the original enzyme; in fact, several different strains were obtained in this way. These new strains have not yet evolved a suitable regulation mechanism. The experiments do, however, show that it is possible for selection acting on new mutations to produce an enzyme with a new function – the ability to break down the 5-C compound. The gene for this enzyme did not of course arise *de novo*, but as a modification of a pre-existing gene. It is also interesting that the new function was achieved in two mutational steps only, and that the intermediate step was also functional. The intermediate enzyme had lost little of its original ability to break down the 2-C and 3-C substrates, but the final enzymes have substantially lower activities on these substrates. If a population is to evolve a new enzyme specificity without losing the old one, this would almost certainly require a prior gene duplication. It will be of great interest to discover the precise changes in amino acid sequence which have taken place in these enzymes.

CHAPTER 6

The Origin and Early Evolution of Life

Figure 13 gives a rough chronology of the earth and of the evolution of life. On the left half of the diagram are given the facts for which there is direct geological evidence; on the right are deductions about the dates of the major events in the history of life which are described later in this chapter. The essential points are these. The earth is a little less than 5,000 million years old. Life probably originated 4,000 million years ago; the first sedimentary rocks are only a little younger than this, and they contain simple bacterium-like cells. Modern nucleated cells, so-called 'eukaryotic' cells, similar to those of protozoa, green algae and all higher plants and animals, appear very much later, being first found in rocks approximately 1,000 million years old. Hence, for three quarters of the period for which life has existed on earth, the only cells were simple non-nucleated 'prokaryotic' cells, found today in bacteria and blue-green algae. The differences between prokaryotic and eukaryotic cells, and the way in which the latter may have originated, are discussed at the end of this chapter.

If we are to discuss the origin of life, we must adopt some definition of living. Elementary textbooks of biology used to contain lists of the defining characteristics of life; the only one I recall is 'irritability', because of the picture it summoned up of an irritable oak tree. Such an arbitrary list is of little use to us. Fortunately Darwin's theory of natural selection provides us with a satisfactory definition. We shall regard as alive any population of entities which has the properties of multiplication, heredity and variation. The justification for this definition is as follows: any population with these properties will evolve by

109

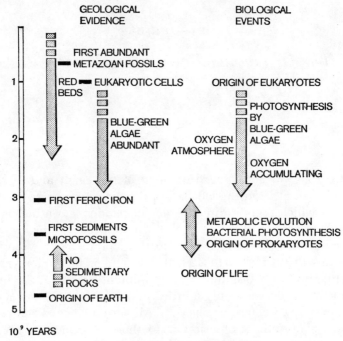

Figure 13. The chronology of life.

natural selection so as to become better adapted to its environment. Given time, any degree of adaptive complexity can be generated by natural selection. The other familiar features of living organisms, in particular their apparent purposiveness, are simply consequences of the primary properties which make evolution by natural selection possible.

The problem of the origin of life, then, is to explain how entities with these properties could originate from non-living matter, without of course invoking natural selection as a cause. If we imagine the simplest conceivable organism whose hereditary mechanism depends on the processes of nucleic acid replication and protein synthesis as we know them from existing organisms, it would have to possess enough DNA to specify all the varieties of tRNA, the protein and RNA components of the ribosomes, the activating enzymes associated with the 20 amino acids, the various enzymes which replicate the DNA and make an RNA transcript of it, and more besides. (It is true that some existing viruses are simpler than this, but only because they rely on their host cell to provide much of the necessary machinery.) It is impossible that an organism of this degree of complexity should arise by physico-chemical processes, without natural selection.

We have therefore two questions. First, what was the nature of the first living organisms, which had a hereditary mechanism simple enough to have arisen without selection? Second, how could the first organisms give rise, by natural selection, to descendants with a fully established genetic code and protein synthetic machinery? We do not know the answers to either of these questions, but we have reached the stage when experimental investigations are making some progress.

(i) *Prebiotic Evolution*

One advantage we have over Darwin is that we have a reasonably accurate idea, derived from radioactive dating and the fossil record, of the age of the earth and of the early history of organisms on it. The earth is approximately 5,000 million years old. In the present context, the most important difference between the earth then and now is that the atmosphere then was a 'reducing' rather than an 'oxidizing' one. To explain this distinction, imagine an atmosphere containing hydrogen (H_2), oxygen (O_2) and water (H_2O). The oxygen and hydrogen would react to make more water. If, when this reaction was completed, excess hydrogen were left over, the atmosphere would be a reducing one; if excess oxygen were left, it would be oxidizing. In a reducing atmosphere, nitrogen is present in the form of ammonia (NH_3) and carbon as methane (CH_4) or carbon monoxide (CO); in an oxidizing atmosphere the corresponding components would be molecular nitrogen (N_2) and carbon dioxide (CO_2).

Our present atmosphere of 80 per cent N_2 and 20 per cent O_2 with traces of CO_2 is a strongly oxidizing one. The reason for thinking that the primitive atmosphere was reducing, or at least that there was no free oxygen, is that in early rocks the iron is deposited in the 'ferrous' form (FeO), whereas in the presence of oxygen iron is deposited as ferric iron (Fe_2O_3), which gives the red-brown colour to rust, and to the soils and sandstones of Devon after which the Devonian period was named.

The importance of a reducing atmosphere lies in the discovery by Miller and others that if a reducing mixture of gases containing oxygen, hydrogen, carbon and nitrogen is treated with high energy, a number of small highly reactive molecules are formed. In the presence of water and ammonia these react to form more complex organic molecules, including sugars, amino acids and nucleotides. On the primitive earth the relevant forms of high energy would have been ultraviolet light, electric discharges from thunderstorms, and high temperatures generated by volcanoes. It has now been shown in the laboratory that almost all the molecules which are the building blocks of proteins and nucleic acids can be formed in this way.

It is presumably not an accident that those organic molecules which are most readily formed play a central role in the bio-chemistry of existing organisms. For example, Oro has shown that if an aqueous solution of ammonia and hydrogen cyanide (one of the readily formed active intermediates mentioned above) is warmed, adenine is formed in substantial quantities. Now adenine is not only one of the four bases of DNA; it is also a component of ATP (adenosine triphosphate), a compound which is fundamental in the storage and exchange of energy in living systems, and of 'cyclic AMP', a compound which appears to play a general role in signalling between cells.

We can now imagine that life originated in what has picturesquely been called the 'primitive soup', in which the necessary molecular components were present in high con-centration. The next stage would have been the formation of strings of molecules, or 'polymers'; in particular, polypeptides (strings of amino acids) and polynucleotides (strings of nucleo-tides). It is not yet quite clear how this happened, because in aqueous solution energy has to be supplied to polymerize either amino acids or nucleotides. It is, however, encouraging that long polypeptides are formed on the surface of a particular type of clay from a solution containing only amino acids and ATP, the latter being the source of energy.

To summarize this section, we can explain how on the primitive earth the important organic molecules (sugars, amino acids, nucleotides) came to be formed, and we may not be too far from understanding how they came to be linked together to form polymers, but this is still a long way from understanding life.

(ii) *Replicating Molecules*

The most plausible conjecture we can make is that the first living things, on the definition given at the start of this chapter, were replicating polynucleotide molecules. Suppose that in the primitive soup polynucleotides acted as templates for the synthesis of further similar molecules. Suppose further that polymers with different sequences of bases existed, and that each type of polymer tended to cause the synthesis of further polymers like itself. Then the properties of multiplication and heredity would be present; since heredity would not be precise, new variations would also arise. It may seem odd to regard such replicating molecules as alive. I will attempt to justify this view in the next section, but first I must discuss whether they could arise.

In existing organisms the replication of DNA requires an energy supply and the presence of specific replicating enzymes. There are two difficulties in imagining how replication might have occurred in the primitive soup, which could not have contained specific enzymes although there may have been random polypeptides. One, already mentioned, is that an energy supply would be needed to link the bases together. The second and more interesting is how the specific nature of a molecule (i.e. its sequence of bases) could be replicated in the absence of a specific enzyme. Some recent experiments by Orgel suggest that it may not be impossible. We will call the relevant building blocks A, U, G, and C (where U represents a molecule similar to thymine; see page 92), and by poly-U we will mean a polymer consisting entirely of U's. Orgel found that if a poly-U molecule were put into a solution of A units, the A units were lined up against it, although they could not be joined together without an input of energy; similarly, poly-C will line up G units, but poly-U has no lining-up effect on G units, and poly-C has no effect on A units. These experiments suggest that replication by complementary base pairing may be possible in the absence of replicating enzymes.

It is a long way from Orgel's experiments to self-replicating nucleic acid molecules, but the objective is now clear. It may be asked why all the emphasis is on nucleic acids and not on proteins. The answer is that in existing organisms nucleic acids replicate but proteins do not. Polypeptides with more or less random sequences of amino acids may well have predated the origin of self-replicating nucleic acids, and may even have helped in the replication process. But, clearly, polypeptides whose amino acid sequence was in some degree determined by polynucleotides (as proteins today are determined by nucleic acids) could not arise before the corresponding polynucleotides.

(iii) *Genotype, Phenotype, and the Origin of the Code*

One of the fundamental distinctions in genetics is between genotype and phenotype. At one level of analysis, this is a distinction between what can be deduced about the genetic constitution of an individual from its ancestry or progeny, and what can be observed of the individual itself. But this depends on a distinction at a physical level between that part of an individual which is directly copied and transmitted to the next generation (mainly, the nuclear DNA), and that part which develops on the instructions of the nuclear DNA and whose function it is to ensure the transmission of copies of that DNA to future generations. The thesis put forward in the last section amounts to the claim that the first living things were naked genes. If so, how and why did the genotype-phenotype distinction evolve?

Any change in a replicating nucleic acid molecule which speeded up its replication would be favoured by selection. We suppose that some polynucleotides had the property of attaching to their surface other small molecules, among which would be amino acids. Some of these attachments would inhibit replication, and the corresponding polynucleotides would be eliminated. Others would favour replication; in these cases, the additional molecules would be the first beginnings of a phenotype.

In some way we have to imagine the evolution of the genetic code, in which particular triplets of bases specify particular amino acids. Initially, coding assignments must have been very imprecise; perhaps a particular group of bases had an affinity for one general class of amino acids rather than another. Perhaps the most difficult problem in the origin of life is to see how a primitive code, which presumably depended on a direct affinity between groups of bases and the corresponding amino acids, could have evolved into the present indirect system with classes of intermediate messenger and transfer molecules and a specialized translation device, the ribosome. We can do little more than ask the questions.

But it is now clear why a distinction between genotype and phenotype had to evolve. The optimum phenotype is determined by external selective constraints; for example, if the survival of an organism requires that it be able to fly, then the shape of its wings will have to meet requirements set by the laws of aerodynamics. A phenotype which meets these requirements will not also be capable of self-replication; a sufficient reason for this is that in a three-dimensional universe it is impossible to replicate a three-dimensional structure by template reproduction. Hence there is a necessary division of labour between that part of an organism which is replicated and that part whose properties ensure survival. This in turn requires a process of development in which the information in the genotype is translated into phenotype. But it may be that at the very beginnings of life this distinction did not exist.

(iv) *The Evolution of Metabolism*

I now want to take a great leap forward in time, and suppose that not only has a modern protein-synthesizing machinery evolved, but that specific enzymes exist catalysing specific reactions, and that the organism has a cell membrane which prevents the products of catalysis from diffusing away. Such an organism could be said to have a metabolism; it would no longer have to depend entirely on organic molecules synthesized abiotically by ultraviolet light or electric discharge. Instead it would be able to make at least some of these molecules from inorganic substances.

Synthesis, however, requires energy. Initially the only source of energy available would have been the amino acids and nucleotides themselves, and other organic compounds which were synthesized abiotically. In a reducing atmosphere these compounds could not have been fully oxidized to obtain energy for synthesis. It is, however, possible to obtain energy from sugars in the absence of free oxygen by a less efficient process known as fermentation; this is still the source of energy for many kinds of bacteria and for animals such as intestinal parasites which live in an environment lacking oxygen. Ultimately the source of energy for life was the sun, but only a small fraction of the sun's energy was converted abiotically into organic compounds. An enormous increase in efficiency, and hence in multiplication rate, resulted when the first organisms trapped the sun's energy directly in photosynthesis.

Plant photosynthesis, carried out by blue-green algae and in the chloroplasts of higher plants, uses solar energy to synthesize sugars from water and carbon dioxide, according to the formula

$$6CO_2 + 6H_2O + solar\ energy \rightarrow C_6H_{12}O_6 + 6O_2.$$

In this series of reactions, the hydrogen required to synthesize sugars is obtained by splitting water, and free molecular oxygen is produced. There is an alternative and probably earlier form of photosynthesis, found today in some bacteria, in which the source of hydrogen is either molecular hydrogen or hydrogen sulphide, and in which no molecular oxygen is produced. This may have been an important stage in the evolution of plant photosynthesis, but was replaced by the former, if only because water is a more readily available raw material than molecular hydrogen.

Photosynthesis was important not only in speeding up the supply of organic compounds to the biosphere, but also in releasing free oxygen. The change from a reducing to an oxidizing atmosphere would have been a gradual one. It is important to understand that the life and death of a plant contributes no net supply of oxygen to the atmosphere. During life the plant synthesizes organic compounds and releases oxygen; after death those same organic compounds will be oxidized, using up exactly the quantity of oxygen which was released in the first place. There will be a net production of oxygen only if dead organisms are fossilized in anaerobic conditions so that their remains cannot be oxidized. In other words, the release of molecular oxygen will be balanced by the production of 'fossil fuels' such as coal, oil and natural gas. Perhaps fortunately, most of the reduced carbon in the rocks is not in readily exploitable form, but in the form of finely divided graphite; consequently when we have squandered our fossil fuels, there may still be some oxygen left in the atmosphere to breathe. There is one other source of atmospheric oxygen whose importance is still in dispute. In the upper atmosphere water molecules are split into hydrogen and oxygen by ultraviolet radiation; the hydrogen escapes into space, leaving free oxygen.

The relative importance of these two processes—photosynthesis plus the formation of fossil fuels, and the splitting of water molecules by ultraviolet light—is uncertain. Between them they gradually converted a reducing into an oxidizing atmosphere. The first evidence of photosynthesis is the appearance of fully oxidized iron, Fe_2O_3, in rocks some 3,000 million years ago; at the same time there is geological evidence for the presence of abundant blue-green algae. It does not follow that the atmosphere at that early stage resembled that of today; more probably there was a gradual shift in the composition of the atmosphere over a period of 1,000 million years or more. Initially, free oxygen would have been as poisonous to primitive organisms as hydrogen cyanide is to existing ones. Ultimately, however, some groups of organisms evolved the capacity to obtain energy from sugars by oxidation instead of by fermentation. Today in higher organisms the process is carried out in special intracellular organelles, the mitochondria, in which sugars and other organic compounds are oxidized to water and carbon dioxide, the energy obtained being used to synthesize molecules of ATP. The ATP molecules can then supply the energy needed for other chemical reactions in the cell.

The presence of free oxygen in the atmosphere had one other important consequence. In the upper atmosphere molecular oxygen, O_2, is converted into ozone, O_3. Ozone absorbs ultraviolet light very effectively, and so prevents these high-energy radiations from reaching the surface of the earth. The early evolution of life took place in the presence of highly damaging concentrations of ultraviolet light. Existing organisms have a number of specific 'repair enzymes' whose function it is to recognize and repair DNA molecules which have been damaged by high-energy radiation or other agents. It seems likely that repair enzymes evolved early, because the need for them would have been much greater before the appearance of an ozone layer in the atmosphere. These enzymes work by cutting out damaged sections of DNA and replacing them by newly synthesized material. Very similar processes occur during the process of chiasma formation and genetic recombination (see page 61) and it is now known that in bacteria some enzymes are involved both in repair and in recombination. It is difficult to say how early in evolution processes of genetic recombination arose; they may have been present almost from the beginning, and have evolved side by side with mechanisms for repairing damaged DNA.

(v) *The Origin of Eukaryotes*

Perhaps the most fundamental distinction between kinds of living organisms is that between 'prokaryotes' and 'eukaryotes' The prokaryotes, which include the bacteria and blue-green algae, have no nucleus or nuclear membrane. The DNA is arranged in a single ring-shaped chromosome with few associated proteins. There are no separate chloroplasts (see page 83) or mitochondria (see page 118). When flagella (simple whiplike structures used in locomotion) are present, they are simple rods made of a single kind of protein subunit known as flagellin. In contrast, the eukaryotes, which include protozoa, green algae, yeasts, fungi and all higher plants and animals, have a nucleus and nuclear membrane. The chromosomal DNA is permanently associated with a variety of proteins. At cell division the chromosomes are attached to a special organelle, the spindle, which plays a role in moving them into the two daughter cells. The cells contain mitochondria and, in algae and higher plants, chloroplasts. Flagella when present are complex structures with a ring of nine peripheral and two central tubules made from a protein known as tubulin.

This is a formidable list of differences and raises problems about how eukaryotes could have arisen. An attractive theory, of which Margulis has been the most persuasive proponent, is that they arose by symbiosis between a variety of prokaryotic organisms: by 'symbiosis' is meant the union in a single functional unit of two or more separately evolved organisms.

Three separate symbiotic events are proposed:

(a) Union between a prokaryotic cell obtaining energy by fermentation, and a prokaryote capable of oxidative metabolism. The latter cell has evolved into the mitochondrion.

(b) Union of the resultant cell with a mobile prokaryote, the entire body of which had filaments arranged in the characteristic $(9+2)$ flagellar structure made of the protein tubulin. It is an important reservation that no such prokaryote is known to exist at present. It is also suggested that the genetic instructions for the eukaryote spindle apparatus are derived from the same source. The reason for this is that the spindle is constructed from tubulin, and shows in some of the details of its structure a remarkable resemblance to a flagellum. The resulting cell would have resembled a simple protozoan; it would be the ancestor of all higher animals and fungi.

(c) Union with a prokaryote resembling existing blue-green algae, which evolved into the chloroplast. The resulting cell could be the ancestor of all nucleated algae and higher plants.

What reasons are there for accepting this view? First, there are a number of existing examples of the symbiotic union of cells which show that the kind of thing proposed does happen.

Perhaps the most remarkable is the protozoan *Myxotricha paradoxa*, which itself lives in the gut of termites. It is a flagellate protozoan which carries with it three kinds of symbiotic bacteria. One of these is a spirochete, a type of motile bacterium; these spirochetes are arranged over the surface of the cell, and they beat synchronously and propel the cell through the medium. This example shows that the theory is not absurd.

The main evidence for the theory is the fact that both mitochondria and chloroplasts have their own DNA and their own protein synthesizing machinery, so that they almost seem to be organisms in their own right. But not all the proteins of these organelles are made *in situ*; some are specified by nuclear genes and synthesized in the cell cytoplasm. So if as most biologists now think, these two organelles were once independent organisms, they have lost their independence during the last 1,000 million years. The evidence that flagellae and spindles were also contributed by a symbiont is less strong but still persuasive.

This chapter has necessarily been speculative. I want to conclude by reviewing very briefly the fossil evidence for this early period. Until recently, it has been something of an embarrassment to evolutionists that the fossil record starts rather abruptly in the lower Cambrian some 600 million years ago, with all the major invertebrate phyla already represented. No fossils were known which were both certainly fossils and certainly pre-Cambrian. This situation is now changing as a result of studies of pre-Cambrian rocks under the light microscope. There is now abundant evidence of fossil bacteria and blue-green algae going back over 3,000 million years. The sudden appearance of the main invertebrate groups 600 million years ago still calls for an explanation. We do now know of an assembly of invertebrate fossils from pre-Cambrian rocks in Australia, 750 million years old, which includes both annelids and echinoderms. These animals were soft-bodied, and are known only because by a rare accident the impression of their bodies was left in the mud. It seems that the real event which took place at the bottom of the Cambrian was the evolution, in a number of different groups of animals, of hard and readily fossilizable calcareous shells. This may have been a response to the origin of the first successful group of invertebrate predators, whose presence would make it selectively advantageous to evolve a protective shell; there is among the early Cambrian fossils a group of arthropods, the Eurypterids, which may well have been these first predators.

CHAPTER 7

The Structure of Chromosomes and the Control of Gene Action

One feature of the evolution of eukaryotes from prokaryotes which is not explained by the symbiosis theory is the origin of the nuclear membrane and complex chromosome structure. It seems likely that these changes were in some way connected with changes in the way in which the activities of genes are controlled. These changes are discussed in this chapter; unfortunately, although we have a fairly clear idea of how genes are controlled in prokaryotes, we are a long way from understanding the process in eukaryotes.

By the 'control of genes', we have in mind the fact that in a cell not all the gene products are manufactured at a uniform rate all the time. In a bacterium, there are three reasons why the rate of manufacture of gene products must vary. Different products are needed:

   (i)   at different times during the cycle of cell division;
   (ii) as the cell finds itself in different environments;
   (iii) when the cell undergoes processes such as encystment or spore formation.

The second of these situations is the best understood, and is the only one I shall describe. As an example, the bacterium *E. coli* does not normally produce the enzymes necessary to break down the sugar lactose. But if the bacterium is grown in the presence of lactose, it will start making the necessary enzymes. It seems therefore that the genes for making these enzymes are there all the time, but are inactive. What keeps them inactive, and what switches them on?

The nature of the switching mechanism is shown in Figure 14.

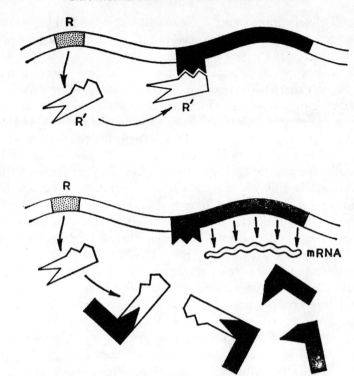

Figure 14. The gene-switching mechanism proposed by Jacob and Monod. Above, a gene, shown in black, is prevented from acting by a repressor substance R′ produced by the gene R. Below, 'inducing' molecules, also shown in black, have entered the cell from outside, and by combining with the repressor substance prevent the repressor from switching the gene off; the gene is therefore producing mRNA.

It was deduced from a study of what happens when it goes wrong; that is, from a study of bacterial strains which cannot be induced to make the enzymes, or which cannot be stopped from doing so. The essential feature is the presence of a 'repressor' gene, R, continuously producing a repressor substance, R′, which moves through the cell to the site of the enzyme-determining genes, and switches them off. In normal circumstances, therefore, no enzymes are produced. But if lactose molecules are present, they combine specifically with the repressor molecules, and so prevent them from switching off the enzyme-determining genes. The mechanism is a kind of double negative; the enzyme is produced not by switching the gene on, but by preventing it being switched off.

In eukaryotes, the capacity to control the activities of genes has been further elaborated. This is particularly obvious during the development of many-celled organisms. During development, the original fertilized egg divides many times, to give rise to many millions of cells, which then differentiate to give rise to nerve cells, muscle cells, bone cells, and so on. For example, during the development of a frog a layer of cells on the dorsal surface rolls up to form a tube, which later becomes the brain and spinal cord. There is a stage, before the tube is formed, when the cells are 'determined'; that is, if they are removed and grafted on to some other part of the embryo, they will nevertheless form a nerve tube. In other words, they differ from other embryonic cells, and transmit this difference to their cellular 'progeny' which arise by division.

It follows that rather stable and long-lasting changes take place during development. Since different types of cell make different kinds of protein, these changes involve the control of gene action. It is known, however, that the changes are not irreversible. Thus King and Briggs were able to take a frog egg, remove the nucleus, and replace it by a nucleus taken from a cell already determined to form part of the nerve tube. They found that such an egg could develop into a normal frog. This shows that all the genes originally present in the fertilized egg are still present after embryonic determination, and that whatever reactions take place between the genes and other cell components during determination are reversible. Nevertheless, the control processes occurring during embryonic development are far more persistent than those occurring during enzyme induction in bacteria.

This more stable type of differentiation in eukaryotes presumably depends on the presence of a nuclear membrane and the more complex chromosome structure. The nuclear membrane separates the two processes of RNA transcription in the nucleus and the translation of the RNA message into protein in the cytoplasm, in this way opening up new possibilities for control. The proteins which are structurally associated with the chromosomal DNA also presumably play some role. This is about all we can say with confidence; but this is an active field of research, and answers are likely to be forthcoming soon.

It is worth saying a little more about the nature and arrangement of the DNA in eukaryotic chromosomes. Our knowledge of this depends largely on the technique of 'molecular hybridization'. The basis of this technique is as follows. If a solution of DNA molecules is heated, the two strands come apart, and one is left with a solution of single-stranded DNA. If the temperature is then lowered, each strand will then 'reanneal' with a complementary strand to form the typical double-stranded molecule; but this can only happen if each strand can find a partner with the appropriate complementary base sequence. The speed at which reannealing takes place depends on the number of 'copies' there are of each sequence. To understand this point, imagine a crowd of 1,000 people (representing single-stranded DNA molecules) who are told that each of them must find one other person with the same surname and then stay with him. Clearly if there are 500 different surnames each represented twice, the rate of pair formation (the rate of reannealing) will be much slower than if there are only 10 surnames each represented 100 times.

If two DNA strands have almost complementary sequences, but have mismatches at a few sites, they will still reanneal, but the hybrid molecule so formed will be less stable when reheated than a molecule with a perfect match. Finally, RNA transcribed from a single-stranded DNA molecule will form a hybrid with it.

By measuring the DNA content of a single human sperm, it is possible to calculate that there is enough DNA to code for well over a million average-sized proteins. This is a startlingly large number. We do not know how many proteins can be determined by the human chromosome set, but most guesses have been of the order of 10,000. If this is right, there is too much DNA by a factor of 100 or more.

One suggestion is that each chromosome is a bundle of identical DNA molecules, so that each cell contains many copies of the same genes. There are two reasons for rejecting this. First, DNA hybridization studies show that although some DNA is present in multiple copies, a substantial part – the so-called unique sequence DNA – is present in only a single copy (or at most in a few copies – the technique is not accurate enough to distinguish). The second reason for rejection comes from a study of mutation. For example, if a male *Drosophila* is irradiated and then immediately mated to a female, it often happens that one of his daughters carries the same new mutation in all her cells; one would not expect this to happen if there had been many copies of each gene on the irradiated chromosome. For these reasons, the idea that a typical chromosome is a bundle of identical DNA molecules can be rejected; to avoid confusion later, I should mention that it *is* thought that the giant chromosomes found in the salivary glands of *Drosophila* are precisely such bundles.

It is possible that a typical chromosome is a single very long DNA molecule, much folded on itself and with protein structures attached to it. Hybridization studies have shown that, in addition to the unique sequence DNA, there are at least two other components. One, the 'highly repetitive' or 're-iterated' DNA, consists of short sequences, only a few bases long, and present in as many as a million copies of each sequence. This DNA differs both in total amount and in sequence between related species, and even between subspecies. A given sequence is distributed in various parts of the chromosome set. It is not transcribed into RNA. No one has any clear idea what its function is. Three possibilities are that it plays a role in chromosome behaviour during cell division, that it is concerned with the folding of the chromosome, or that it has no function at all. It certainly changes very rapidly, so that there must be some means whereby a particular sequence can be replicated many times and distributed throughout the chromosomes. I mention it not because it helps to explain anything, but because it is an outstanding puzzle.

A second DNA component consists of much longer sequences which are repeated some hundreds or thousands of times in one cell. At least some of this DNA is transcribed into RNA. Again its function is somewhat obscure, although we can account for a small part of it. Thus it is known that the gene which codes for the RNA of ribosomes is present in a number of copies, presumably because a large amount of the RNA transcript is needed in each cell. The same is true of a few other genes.

Even when the repetitive DNA has been allowed for, the unique sequence DNA still amounts to some half of the total. Thus we are still faced with the problem of too much DNA. An explanation is beginning to emerge from a study of *Drosophila*. It seems that there are only some 5,000 'structural genes' in *Drosophila*; the term structural gene means a gene which is translated into a protein product. A structural gene will be in the range 300–3,000 bases long, depending on the size of the protein it specifies. Associated with each structural gene is a much longer stretch of DNA which is not translated into protein but which is transcribed into RNA. Presumably this RNA is important for control. One way in which it might be so is as follows. A long RNA molecule may fold up in a specific way; this it does by 'doubling back' on itself if the base sequence in one part is complementary to the base sequence in another part of the same molecule. Thus an RNA molecule can not only carry a genetic message; it also has a 'phenotype' of its own. Hence each messenger molecule when it is first produced will be attached to a much larger RNA molecule whose folding pattern may determine whether the messenger is translated.

Of course this is only one possibility, and a speculative one at present. What is now reasonably certain is that only a small fraction of the chromosomal DNA of higher organisms consists of unique sequences which are translated into proteins – the so-called structural genes; the remainder, some of which are highly repetitive in sequence, perform a variety of functions, most of which remain to be discovered.

Despite the uncertainty as to how DNA molecules are arranged in chromosomes, quite a lot is known about the way in which major changes in chromosome structure originate. Much of this knowledge depends on the fact that during meiosis 'homologous' chromosomes pair; by homologous chromosomes are meant those which carry identical, or at least very similar, sets of genes. Hence a study of the pairing of chromosomes in hybrids between different species or races provides a method of establishing the common descent of chromosome segments in the two species.

In *Drosophila* and in other two-winged flies this method can be further extended, because homologous chromosomes pair not only in meiosis but also in the salivary gland cells, in which the chromosomes are of great size. Consequently pairing can be studied in great detail, including pairing in hybrids which fail to form proper ovaries or testes.

Each salivary 'chromosome' is in fact a bundle of similar chromosomes, and characteristic bands can be seen across them; it now seems that each band may correspond to a single structural gene. It is possible to distinguish two kinds of change in chromosome structure, the first involving changes in the arrangement of a given quantity of chromosome material, and the other changes in the total quantity of such material.

The two commonest kinds of structural rearrangement are 'inversions' and 'translocations'. If, either spontaneously or after irradiation or treatment with mutagenic chemicals, a chromosome is broken, the broken ends will usually rejoin. If two breaks occur simultaneously in the same chromosome, the middle piece may rejoin the end pieces in an inverted position. Such an event is detected in the following way. If a chromosome carrying such an inversion passes to a gamete, and unites in a zygote with an uninverted chromosome, the two will pair by forming a loop, as shown in Figure 15. Such inversion loops, which are easily seen in salivary gland nuclei, are the natural consequence if each region of a chromosome pairs with the homologous region of its partner. It is one of the most remarkable facts in the history of genetics that both the existence of inversions and their mode of pairing were predicted by Sturtevant from purely genetic evidence, before there was any cytological evidence on either point.

Such inversion loops were found in hybrids between *Drosophila melanogaster* and the closely related species *D. simulans* by Patau and by Kerkis, and have since been seen in many other species hybrids. They are, however, common within many

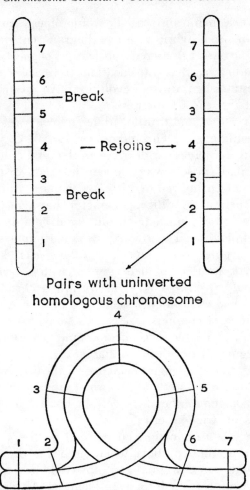

Figure 15. Diagrams of the origin and mode of pairing of an 'inverted' chromosome.

Drosophila species, a fact which has been exploited in the study of natural selection in a way which will be described in Chapter 10.

Should breaks occur simultaneously in two non-homologous chromosomes, it is possible for the four fragments to rejoin after an exchange of partners, to produce a translocation. The pairing of two such translocated chromosomes with two normal chromosomes in a hybrid is shown diagrammatically in Figure 16. Such patterns are a common feature of meiosis in species hybrids, indicating that translocations have occurred and have become established during evolution. An example will be discussed further in Chapter 16.

Both inversions and translocations involve a rearrangement of a given quantity of chromosome material. Other changes, resulting in an increase or decrease in the total amount of such material, may arise owing to an unequal distribution of chromosomes during cell division, particularly during meiosis. In experimental conditions, gametes have been observed carrying twice the normal chromosome complement, i.e. a complete diploid set. The union of two such gametes would give rise to a zygote with four chromosome sets, known as a tetraploid, or the union of such a gamete with a normal, haploid gamete would give rise to a zygote with three chromosome sets, a triploid. The evolutionary relevance of such events will be discussed in later chapters.

Finally, it is possible for a chromosome to arise which either lacks a segment (deletion) or has a segment present twice (duplication). The strongest evidence for the occurrence of such events comes from a study of abnormal salivary gland chromosomes, which may either lack certain bands, or have certain bands present twice in series.

A process of duplication could lead to the type of gene duplication postulated above when discussing the evolution of haemoglobins; later, original and copy could be separated on to different chromosomes by a translocation. Alternatively, the incorporation of an extra copy of a whole chromosome into the normal chromosome set could in one step duplicate the genes on a chromosome and separate the copy from its original.

It is interesting that chromosome inversions are so often compatible with life. Thus it was pointed out (page 97) that the inversion of a long segment of a DNA molecule would necessarily have disastrous results. How is this to be reconciled with the suggestion that the DNA of a chromosome may be arranged in a single long molecule? Two things seem to be

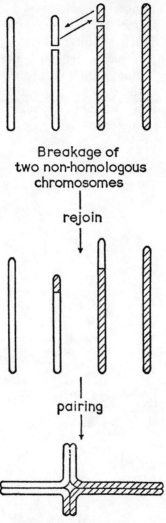

Figure 16. Diagrams of the origin and mode of pairing of 'translocated' chromosomes. One pair of homologous chromosomes, and segments which are copies of them, are shown hatched.

required. First, if an inversion is to be viable, the breaks must occur in regions which are not carrying essential genetic information; now that we know that there is much repetitive DNA this is not unreasonable. Second, although it will always be the case that only one of the two strands of DNA carries a meaningful message and hence is transcribed, the choice of a strand for transcription must switch from strand to strand as one travels along the chromosome.

CHAPTER 8

Variation

We take it for granted that we can recognize individually our friends and acquaintances. This is possible only because no two of them are alike; there are sufficient differences of shape, movement, and voice to distinguish them. It is less obvious, but equally true, that such individual differences exist between animals. They are more difficult for us to recognize because we are less accustomed to looking closely at animals than at human beings. Similar difficulties of recognition may arise between different human races. To Europeans, Chinese may all look alike at first acquaintance; a recent delegation of Englishmen to China found the Chinese could not easily distinguish the members of the party. Such difficulties soon disappear on closer acquaintance with other racial groups.

In the same way, human beings can learn to recognize wild animals individually. Miss Len Howard, who shares her cottage with several species of garden birds, can recognize individually a number of Great and Blue Tits. She finds that there are differences, not only of plumage and shape, but of movement, temperament, and aptitude between members of a species. There is nothing 'unscientific' about this kind of observation. Although it is one of the jobs of zoologists to make statements which are true of all Great Tits, or for that matter of all birds, it is important to study the differences as well as the similarities between the members of a species; in the absence of such differences evolution would be impossible. It is interesting that, at least in many vertebrate species, animals can and do recognize one another as individuals. For example, Tinbergen found that a Herring Gull is able to recognize its mate, both by its voice, and also by sight, even when flying thirty yards away. A gull can also recognize birds which occupy neighbouring territories on the breeding grounds, and can recognize its own chicks a few days after hatching. In none of these cases does recognition depend on an individual being in a particular territory; the gull is recognizing the voice and appearance characteristic of an individual.

In human beings, recognition depends primarily on differences of shape and of voice. It is worth mentioning some other respects in which no two individuals are alike. The police have found that differences in a character as trivial as the pattern of ridges on the fingers are sufficient uniquely to identify a human being. Differences also exist in the antigens carried on the red blood corpuscles, which determine a person's 'blood group'. These concern not only the well-known *ABO* antigens and the *Rhesus* factors, but also a number of other antigen systems. The number of such differences at present recognizable is insufficient to enable us to distinguish any two individuals, but the chance that two individuals will resemble one another for all known blood groups is only about 1 in 100. However, an almost complete discrimination between individuals is possible by considering skin antigens. Skin from one part of the body can be grafted on to another part of the same individual, where it will grow and establish itself. However, skin from one individual grafted on to another will induce an 'immune reaction' as a result of which it is sloughed off. Grafting, therefore, is a very sensitive method of detecting biochemical differences between the skin of different individuals. Both blood and skin antigens are known to be genetically determined. Consequently an exception to the rule that skin from one individual cannot be grafted on to another concerns monovular twins, i.e., twins who have developed from the same fertilized egg, which has divided into two at an early stage of development. Conversely, a recent and much publicized claim of 'virgin birth' was effectively disproved when it was found that a graft of the child's skin on the mother was thrown off; had the claim been true, the child could not have possessed antigens which the mother lacked, and consequently no immune reaction could have occurred.

Like differences in finger-prints, individual differences in smell are also used by the police. It has been found that dogs are able to follow the track of an individual after being presented with clothing belonging to that individual. This ability has recently been investigated by Kalmus, who found that dogs were able to recognize the smell of an individual human being, even when the smell was mixed with that of other people. The dogs found greater difficulty in distinguishing monovular twins, but even this was possible in some circumstances.

Most studies of variation and of natural selection have been concerned with more striking and easily recognizable differences. In such studies, it is important to distinguish between differences between individuals inhabiting a given region, and differences between populations from different regions. Differences between members of a single population are an example of 'polymorphism'; however, this term is usually confined to cases where two or more sharply distinct forms exist in a population. Thus the human population of Great Britain is polymorphic for the *ABO* antigens; individuals may be *A*, *B*, *AB*, or *O*, and intermediates do not exist. It is also highly variable in stature, but this would not be regarded as an example of polymorphism, since no sharp distinction can be made between 'tall' and 'short' individuals. These are examples of two extreme types of variation, between which there are many intermediate patterns. Thus for human eye colour, there are people with blue eyes and others with brown eyes, but there are many others with intermediate shades of eye colour.

The *ABO* antigens are determined by alleles at a single locus, and are unaffected by environmental conditions; stature is influenced by genes at many loci, and also by nutrition. This will explain why in this book, and particularly in discussing selection in wild populations, more examples of polymorphism will be described than of continuous variation. This is not because polymorphism is commoner, or of greater evolutionary importance, but because it is easier to study. This is a particular example of a more general fact which is familiar to professional biologists, but which may not be obvious to laymen. The examples given in this and similar books were originally chosen for study because a particular animal or plant offers special advantages for the study of a particular problem; genetic studies are made on *Drosophila* because it breeds quickly, does not suffer from infectious diseases, and has giant chromosomes in the salivary glands; study of the behaviour, taxonomy, and distribution of birds is more advanced than that of most other groups because they are large, conspicuous, diurnal animals, and so on. There are of course dangers in building up our picture of evolution from a series of examples selected in this way. The picture may prove to be a distorted one, but it is better than no picture at all.

Some of the most striking examples of polymorphism among English wild animals occur among the butterflies and moths. Most females of the Silver-washed Fritillary, *Argynnis paphia*, resemble the males in having a rich brown ground colour on the wings, with dark markings. However, in the New Forest from 5 to 15 per cent of the females are a dark olive-green colour; elsewhere such females are rare. In the Scarlet Tiger-moth, *Panaxia dominula*, the normal form, *dominula* has dark fore-wings with a number of pale spots, and red hind-wings with dark markings. There is also a dark variety, *bimacula*, with only two pale spots on the fore-wings, and more extensive dark areas on the hind-wings, and there is an intermediate variety, *medionigra*. These darker forms are rare in most localities, but were found to be quite common in a marsh near Oxford. Perhaps the most surprising fact about this latter example of polymorphism, discovered by Sheppard, is that females prefer to mate with males of a different colour from their own. In both *Argynnis* and *Panaxia*, the differences in colour are due to gene differences at a single locus.

Returning to our own species, in addition to the differences between individuals inhabiting a given region, there are also differences between populations from different areas. Human races differ in the colour of their skins, in the colour and structure of their hair, in the shape of their skulls, in the frequency of occurrence of different blood groups, in their resistance to various diseases, and in many other ways. These differences are either wholly or mainly determined at birth, and are not the result of differences in climate or nutrition. There are other differences between human races which are not so determined, but which are the result of the conditions in which each individual grows up. The language which people speak, and the religious beliefs which they hold, are largely a matter of custom and tradition.

A difference between two populations may be an absolute one, with little or no overlapping between the two groups, or it may be only a matter of average values. The difference in skin colour between Englishmen and Africans is an absolute one, or almost so; nearly all Negroes have darker skin than nearly all Englishmen. On the other hand, while it is true that Italians have on the average darker skins than Englishmen, there are plenty of exceptions to this rule, so that it would be impossible to tell from the colour of a man's skin, or indeed from any other purely physical measurements or observations, whether he was Italian or English, although a guess would have a fair chance of being correct.

Just as in the case of individual differences, many people who are well aware of the differences between human races do not appreciate that similar geographical variation occurs in other animal and plant species. In fact, in every animal species which has a sufficiently wide geographical range, and which has been adequately studied, differences have been found between the populations inhabiting different regions. This fact first became apparent to taxonomists in the study of birds, but has since proved to be true of other groups. It is an illuminating exercise to glance at the sections headed 'Characters and Allied Forms' in Witherby's *Handbook of British Birds*. For almost every bird species on the British list, a number of other geographical races are mentioned, differing in details of plumage or beak shape – the only characters easily available to the taxonomists. Occasionally you will come across the phrase 'doubtfully separable', implying that the difference between two populations is a statistical, not an absolute one.

A particularly interesting example of geographical variation in birds is provided by the Carrion and Hooded Crows. The Carrion Crow, the common crow of England and southern

Scotland, is black, whereas the Hooded Crow of northern Scotland has a light-grey mantle. In Witherby's *Handbook* the two forms are listed as separate species, but since they interbreed readily where their areas of distribution meet, it is perhaps better to regard them as different races of the same species.

The Hooded Crow inhabits the whole of central and eastern Europe, Scandinavia, Russia, and the western parts of Siberia. In the east its range meets that of the Carrion Crows of eastern Siberia in a zone of interbreeding running from the mouth of the Yenisei south to the Altai mountains, and thence south-west to the Aral Sea. A similar zone of interbreeding, running from the Baltic to the Gulf of Genoa, divides the Hooded Crows from the Carrion Crows of western Europe, and a third such zone crosses central Scotland. Such a distribution in present-day Europe seems to make little sense, and must be explained in terms of the past. On an evolutionary time scale, the recent past saw central and northern Europe covered by the ice-sheets of the last Ice Age. As the ice spread southwards, the crows must have retreated before it, until they were split into isolated populations in south-western Europe, in south-eastern Europe and the Middle East, and in eastern Siberia. If at this period the central of these three groups evolved the hooded pattern, then their northwards spread after the Ice Age would give a distribution such as we find today. The Hooded Crows have probably reached Scotland from Scandinavia. This explanation is, of course, guesswork, in the sense that we have no direct evidence as to the distribution of the two forms at the end of the last Ice Age, but given the present distribution and our knowledge of the spread of the ice-sheets, it is probably the correct one.

However, it is likely that the Hooded Crows are better adapted to the colder regions of central and northern Europe, and the Carrion Crows to more temperate regions. If this were not so, it is difficult to see why the intermediate hybrids should be confined to relatively narrow zones, some hundred miles in width. We would rather expect to find that the movements of individual birds led to the intermingling of the two populations to form a single, possibly rather variable population. The maintenance of a narrow zone of hybridization of this kind is favoured if the range of movement of individual animals is restricted, and if there is fairly intense selection in favour of one form on one side of the boundary and of the other form on the other side.

The present diversity of human races probably has similar though much more complex origins. In the early Stone Age man must have been a relatively rare animal, divided into a number of more or less isolated populations. These populations diverged in the colour of the skin, the texture of their hair, and in many other ways, partly in adaptation to local conditions and partly fortuitously. With the invention of agriculture and the consequent increase in human numbers, and with improved methods of travel, these populations have again come into contact with one another. From this point on, however, the history of man seems likely to be different from that of the crows. Already much intermingling has occurred between once distinct populations, and there is little doubt that this process will continue, despite the existing political and cultural barriers to intermarriage

Granted that all widely distributed species are found to show geographical variation, how far do the differences result from adaptation to local conditions? Unfortunately this question is difficult to answer. For reasons of expediency, most taxonomic studies are carried out on dead specimens – often, in the case of mammals and birds, on skins. One of the important jobs of a taxonomist is to provide descriptions which will enable other biologists readily to identify specimens which they may collect. For this purpose, descriptions of the plumage of a bird are convenient, whereas descriptions of its internal anatomy, its behaviour, or its physiology are not, even though the latter features might be a better guide to its adaptation to a particular locality. In recent years there has been an increasing tendency for taxonomists to study many different aspects of the groups they describe, but it remains true that it is usually more difficult to recognize the adaptive significance, if any, of the relatively minor differences between geographical races of a species than of the most striking differences between species or genera.

To the question whether or not a particular difference

between two geographical races is or is not adaptive, three kinds of answer are possible:

(i) That the difference is indeed adaptive. For example, the different degrees of skin pigmentation in the inhabitants of Africa and of northern Europe are surely adaptations to the intensity of sunlight. Similar examples can be given from animals. It was mentioned in Chapter 1 that the number of eggs laid by robins from high latitudes is greater than by robins breeding nearer the equator, and suggested that this may be an adaptation to the hours of daylight during the breeding season. In many species of mammals, the individuals from colder regions are larger, but have smaller ears and shorter tails. In a warm-blooded animal these features will tend to reduce the rate of loss of heat from the surface. The physiological adaptations of northern and southern races of the American frog *Rana pipiens* will be described in some detail in Chapter 15.

Similar 'clines' of continuous variation within a species are common among plants, the term cline implying a more or less continuous change in the characters of a population in space. In *Pinus sylvestris*, Langlet found variations in chlorophyll content, length of leaves, hardiness, and rapidity of shoot development in spring in pines from different latitudes. Clausen, Keck, and Hiesey observed clines in the height of *Achillea lanulosa* (a relative of Milfoil) at different heights above sea level, the taller plants growing at lower altitudes. Such a cline is commonly found where the variation in external conditions is also continuous; a different type of geographical variation was described by Turessen in *Hieracium umbellatum* (Hawkweed). In a small region of southern Sweden, he found populations adapted to sand dunes, to fields, to cliffs, and to woodland. Such locally adapted forms he called 'ecotypes', a term which implies that the differences are genetically determined, and not a response of each individual during growth in particular conditions. He was able to show that the dune ecotype had arisen independently a number of times, since, although in all cases adapted to life on shifting sands, the ecotypes from different dunes tended in other ways to resemble the inland forms from the same area.

(ii) That the differences are purely fortuitous, and not adaptive.

(iii) That the observed differences are not adaptive, but that they are associated in development with other, unrecognized differences which are adaptive. For example, in a number of warm- blooded species, the coloration tends to be darker in hot and wet climates, and lighter in cold and damp climates. It is far from clear why such differences should be adaptive, and tempting to suggest that they are by-products of some other physiological adaptation. As pointed out in Chapter 3, two superficially unrelated differences may be produced by a single genetic change; for example, grey fur and a particular skeletal abnormality in mice. It is therefore quite possible that the differences observed by a taxonomist may similarly be related to other differences which he cannot see.

There are, however, good reasons why such an explanation should not easily be accepted. It is in essence an act of faith, not a statement of observed fact. Unless it is followed by an attempt to discover the underlying adaptive differences, and to dem- onstrate the connection in development between these and the supposedly non-adaptive ones, the significance of geographical variation is left unexplained. There is in fact no case in which such an explanation of geographic variation has been proved to be true, and there are reasons for doubting whether such a process would in fact occur. Let us suppose that some adaptive change, A, is associated on its first appearance in evolution with some other character, B. The character B is unlikely to have no effect of fitness in the environment which favours A. If B also increases fitness in this environment, it would be expected to spread anyway, and there is no need to invoke the existence of A to explain its appearance in the population. If the associated character B has an adverse effect on fitness, then it is probable that in time new genetic variations will arise to break the association between the two characters, so that the favourable character A can spread without involving also the appearance of character B.

We must now consider how far the differences between geographical races are genetically determined, and how far they are due to the direct action of the environment during the lifetime of individuals, as are the languages and religious customs of human races. This problem can only be solved by experiments, which are most readily carried out on plants. It is fairly easy to take seeds or cuttings of a wild plant and grow them in the conditions normal for a different race. Such experiments have usually shown that the differences are at least in part genetic in origin. For example, Figure 17 shows the heights of *Achillea lanulosa* plants from four different regions of California, when grown side by side in the same garden. The differences in height are similar to those between the parent plants grown in their native habitats, but less in extent. Therefore the differences in this case are partly genetic and partly environmental in origin.

The same conclusions probably hold for animals, but the experiments are more difficult to perform. Conditions in zoos and laboratories are abnormal and animals raised in them often differ, not only from members of the population from which they came, but also from all other wild populations. If however, wild animals from one locality are released in another, it may be difficult to recover them or their offspring in order to determine the result of the experiment. One such successful experiment, on the migrating behaviour of storks, was described in the first chapter and showed the differences to be inherited.

However, even when the difference between two races is genetic, as is certainly the case for the dark skin of Africans and probably so for the small ears and short tails of animals living in colder climates, a similar result may be caused by the developmental flexibility of individuals. Europeans living in the tropics develop a darker skin, and laboratory mice raised at low temperatures have shorter tails.

So far in this chapter we have been concerned with variation in wild populations. Far more extreme forms of variation are observed in domestic animals. It may seem puzzling at first sight that animals as different as a dachshund, a bulldog, and a whippet are placed in the same species, *Canis familiaris*, whereas the wolf, *Canis lupus*, and fox, *Vulpes vulpes*, although superficially resembling one another more closely than do a bulldog and a

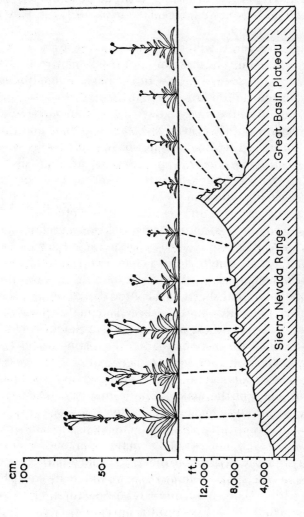

Figure 17. Variation in typical plants of *Achillea lanulosa*, all grown in the same garden, but originating at different heights above sea level, in the localities shown in the transect of California (after Clausen, Keck, and Hiesey).

whippet, are in different genera. Before explaining the reasons for this, I shall describe first another animal whose domestication is much more recent, although it extends over some 400 generations, and much better documented, namely the fruitfly *Drosophila subobscura* kept in the genetics laboratory at University College London.

In addition to several 'wild-type' stocks, there are some fifty 'mutant' stocks. Many of these mutant stocks can readily be distinguished from the wild-type with the naked eye, although the fly is only a few millimetres in length. Such stocks may have white eyes instead of the dark-red eyes of wild flies, they may have a pale yellow cuticle instead of a dark grey one, and their wings may be curled, twisted, or misshapen in various ways. Recently a stock of flies with only four legs, instead of the six legs typical of insects, was established. It so happens that the first four-legged flies were found in a stock already characterized by the possession of legs with an excessive number of joints in the tarsal segments of the legs, and by wings held out at right angles to the body instead of being folded over the back. Each of these striking differences is due to a gene change, or mutation, at a single locus; in fact, a new stock is often established by breeding from a single individual showing the altered character.

From time to time there have also been kept in this laboratory a number of wild-type stocks of other closely related species of *Drosophila*. Some of these species cannot be distinguished with the naked eye, and may be so similar in general appearance that a study of such details as the bristles on the ovipositor or the shape of the plates of the male genitalia may be necessary in order to determine to which species a specimen belongs. Yet there is no doubt of their specific distinctness, since members of different species will not normally mate with one another. Significantly, the simplest way to make sure that a particular fly does belong to the species *subobscura* is to watch its courtship behaviour with a known member of the species, since this behaviour is strikingly different in the various species.

Thus we have a contrast between, on the one hand, very simple genetic differences, involving single loci, which produce large and easily recognizable changes in appearance, and, on the other, very complex genetic differences between species producing far more subtle and less obvious changes. Clearly not all simple genetic changes at single loci can produce such gross results as did those which formed the origin of the mutant stocks; the genetic differences between species are probably built up in the main of gene changes at a large number of loci, each individual change producing results which would not easily be noticed. But only those genetic changes which did produce easily recognizable results were noticed in the laboratory, and hence only they could be used to establish a new stock.

Returning to the more familiar domestic dogs, it is generally agreed that they are descended from wolves, although other species of wild dogs, for example jackals, have probably contributed to their ancestry. Hybrids between domestic dogs and wolves are easily obtained and fully fertile. Mating between dog and vixen, or reciprocally, has often been attempted but never observed, and claims that fox-dog hybrids have been obtained are not generally accepted.

There is an obvious parallel between the striking differences between breeds of dogs, and between the mutant stocks of *Drosophila*. A number of the most striking breed characters are in fact due to single genes. Thus the hanging ears of spaniels are dominant over the 'wild-type' upright ears, a short tail is dominant to the normal long tail, and the short legs of dachshunds and sealyhams are dominant to the normal long legs. Similarly most of the variations in the colour and texture of the fur are due to relatively few genes showing simple Mendelian inheritance. However, it would be wrong to assume that the differences between the longer-established breeds is a matter of only one or a few genes. Rather, having been given a starting-point by some major mutation, men have further elaborated and refined the breed by selection of many other modifying factors. Nevertheless, the extent of the genetic difference so produced is still small when compared with the difference between species, if we are to judge by the complete fertility of most interbreed hybrids.

It is now possible to see why it is that a much wider range of apparent variation has been produced in domestic species than in wild ones. A domesticated population has been largely removed from the action of natural selection. Consider, for example, the question of colour. Except in the Arctic, a white mammal would be at a serious disadvantage, whether in hunting or in escaping from predators, and in fact white individuals are very rare in the wild. Yet there are plenty of tame white mice, rabbits, rats, dogs, and cats. Further, white mammals in the Arctic are always black-eyed whites, whereas many of the white forms of domestic animals are pink-eyed albinos, and almost blind.

The range of variation is often more extreme in those domestic animals such as mice, pigeons, or dogs, which are widely bred by fanciers, than in those such as cattle, pigs, and horses, which have an economic value in agriculture. In the former case the breeders can choose their own objectives and abnormal characters such as waltzing in mice or tumbling in pigeons, can be established in a breed. In the latter group, the aims of the breeder are set for him by the economic needs of society – at the present time by the profits to be made out of a particular breed in particular conditions of agriculture. Hence much of the variation in economically important breeds arises because the conditions of agriculture are not everywhere alike, and because a given species of domestic animal may be exploited for several purposes. Thus highland cattle, zebu cattle, and Jersey cattle are all adapted to particular conditions, and, at least in the more advanced countries, specialized breeds have been developed, as in the case of milk and beef cattle, racehorses and cart-horses, or pork and lard pigs.

The wide variety of dog breeds was originally a response to the number of different animals which can be hunted with the aid of dogs and the variety of methods of hunting, although today dogs are bred more for show than for use. Greyhounds and beagles are both intended for the hunting of hares, but one has been selected to run as fast, and the other as slowly, as possible. The dominant short-legged condition has been bred into beagles to reduce their speed and into dachshunds and sealyhams to enable them to hunt in burrows. Not only structural peculiarities, but also inherited peculiarities of behaviour have been made use of. All dogs differ from their wolf ancestors in two features of behaviour: they bark and they wag their tails. Neither of these acts is normal to wolves, though both can be taught. Behavioural differences between breeds include the habit of 'pointing' at the scent of game in contrast to pursuit, and the habit of baying when hunting in contrast to silent hunting. These differences are in part inherited and in part the result of training.

Enough has been said to show that much of the variation found in domestic animals has arisen because they have been partly removed from the influence of natural selection and because they have been divided into a number of populations, or breeds, selected for different purposes. The process is a reversible one; dingoes, wild descendants of domestic dogs taken to Australia by man, are no more variable than other wild species.

It would, however, be wrong to assume that all the changes in domestic animals are the intended results of conscious human intervention, Spurway has pointed out that the founder members of any domestic population must consist of those individuals, often abnormal, which can be tamed and which will breed in captivity. Further, the process of domestication must change the breeding system, resulting sometimes in hybridization between different geographical races or even species, and sometimes in mating together close relatives. In either case the consequence may be a large and often unexpected change in the nature of the population.

It is also true that some form of natural selection influences even domestic populations. An economically successful breed must be adapted to local conditions. An anecdote by Hagedoorn will illustrate this point. In northern Holland, small black ducks are caged during the day, and let out in the evening. They feed during the night along the canals, returning at dawn to lay eggs during the day. The particular adaptability of these ducks to local conditions was not fully appreciated until an attempt was made to introduce in their place large white Indian Runner ducks. Since the ducks did some damage to the banks of the canals, many complaints were made about the conspicuous and diurnal white ducks which had not been provoked by the more discreet and nocturnal black ones, and a number of white ducks failed to return. The same principle is true, on a larger scale, of the resistance to disease and climatic conditions of local breeds, and of their productivity on the particular types of pasture available to them.

In addition to the relaxation of natural selection and the division of domesticated species into breeds artificially selected for various purposes, there is another reason why such a large range of variation can be produced in such a relatively short time. In domestic animals, men can choose not only which individuals are to be the parents of the next generation, but also which individuals are to be mated with which. The importance of this can be illustrated by considering the method of establishing a stock of albino mice. The starting-point must be the existence of one or several albino mice. Today, of course, such albinos could be bought in any pet shop, but in the absence of any such source of supply, there is no known method of producing albino mutants 'to order', although, as explained earlier, there are many ways of increasing the frequency of all kinds of mutations. However, such albinos would occur spontaneously, although with a very low frequency. Albinism in the mouse is due to a recessive gene, c. Once an albino mouse was obtained, it would normally be mated to a wild-type mouse, since no other would be available. Since the gene is recessive, the F_1 progeny would be wild-type. However, if these were mated together, one quarter of the F_2 would be albinos, and if the albinos were mated together, in the third and subsequent generations a stock would be obtained in which all the mice were albino.

The critical point here is that the F_1 animals should be mated together, and that in later generations albinos should be mated with albinos. Now consider what would happen if albino animals, when they occurred, were bred from, but if the pairing of animals was left to chance in a large population of mice, as would be the case in nature. The original albino individual would be preserved as a parent, so there would be in the next generation some mice heterozygous for the albino gene c. However, such mice would be rare compared to mice homozygous for the wild type allele of c, and therefore would be likely to mate with such homozygotes. If so, no albino mice would appear in the second generation and it might be a considerable time before another albino mouse was born.

Thus the ability to decide which individuals are to be paired with which makes it possible quickly to establish stocks breeding true for characters which otherwise would remain as rare variants or 'sports' in the population. Since the most important rule in any such programme of selective mating is the mating of like with like, the process was effective long before the discovery of Mendelian inheritance. There is, however, one domestic animal for which human beings only rarely decide upon which individuals are to be paired together, namely the domestic cat. It is no accident that cats are far less variable in size and shape than are dogs, although they do show some variation in colour and length of fur.

The fact that mating in wild populations is, at least to a first approximation, at random has been an important stabilizing factor in evolution. Even where a population develops characters, differing from those of the rest of the species, which adapt it to local conditions, it is always liable to be 'swamped' through migration and interbreeding. The problem of the origin of new species, discussed in later chapters, is to a large extent the problem of how this levelling-out effect of interbreeding has been overcome in the wild.

CHAPTER 9

Artificial Selection: Some Experiments with Fruitflies

In the light of what has been said in the preceding chapters about variation and about heredity, we can now return to the question of how far natural selection as postulated by Darwin can explain the known facts of evolution. I shall first describe some results of artificial selection in the laboratory and then consider the evidence that natural selection is acting on wild populations at the present time. In both these situations the time involved is short, covering only tens or at the most hundreds of generations, and the changes occurring correspondingly small. In later chapters I shall have to try to bridge the gap between such short-term changes of slight extent and the more profound changes which have taken place during the hundreds of millions of years during which organic evolution has proceeded.

I want now to describe the effects of selection in altering characters which are influenced by many pairs of alleles at many different loci. It was stated earlier that for such characters we should expect to get a distribution in the population similar to that in Figure 5. That statement must now be justified, but first there is a difficulty to be discussed. Differences in continuously varying characters, for example stature, may not be genetic in origin at all, but due to environmental conditions. How can we be sure that all the differences in stature shown in Figure 5 are not due to differences of nutrition or other environmental conditions?

The main reason for thinking that at least some differences in stature are genetic is that close relatives tend to resemble one another. A man is more likely to resemble his father or his brother in height than to resemble another man taken at random from the population. Of course, this does not prove that similarities between relatives are genetically determined, since members of the same family usually share a common social background and standard of life. However, even in a group of people drawn as far as possible from the same social class, we still find closer resemblances between relatives than between unrelated individuals. When studying laboratory animals, it is possible to standardize the environment still further until most of the observed variation is due to genetic causes.

The method of investigating the inheritance of a character such as size in a laboratory animal is roughly as follows. First, certain assumptions, or guesses, are made about the number of different genes involved and the ways in which they will influence size. From these assumptions, it is then possible to calculate, for example, the amount of resemblance to be expected between parents and offspring, or between sibs (offspring of the same parents), or the results to be expected from selecting for larger or smaller size, or from inbreeding (i.e. mating together in each generation close relatives, for example brothers and sisters). These expected results, obtained by calculation from the original assumptions, can then be compared with the results of actual experiments. If the calculated and observed results do not agree, then either the calculations were wrong, which is unlikely if reasonable care was taken, or one or more of the original assumptions were wrong, which is only too likely. The kind of difference between observation and calculation will often show which of the original assumptions was erroneous.

Of course, the various steps are not necessarily made in this order; the experiments may be performed first, and an attempt to find assumptions to explain the results may follow. The essential point is that an attempt is made to devise a model, involving assumptions about the actions and interactions of the genes concerned, the influence of environmental conditions, and perhaps of maternal and cytoplasmic factors, which will account for the observed results and from which further predictions can be made.

This method of approach will now be illustrated by an example. The animal concerned is *Drosophila melanogaster*, which is convenient to study, since many generations can be obtained each year, and because a single pair may have several hundred offspring. Adult flies vary in the number of bristles on the abdomen, a character which sounds, and probably is, trivial, but which has been popular with geneticists because it gives them something to count, and because bristle number is not greatly influenced by environmental conditions. The results which have been gained from the study of bristle number are important, although as we shall see later they may sometimes be misleading if applied uncritically to such problems as selection for economically important characters in domestic animals, or natural selection in wild populations.

In actual investigations, the investigator spends much of his time worrying about the effects of environmental conditions. In the model, it will be assumed that all variation in bristle number is genetic; this is more nearly true of bristle number than of many other characters and the assumption will make the argument easier to follow. It will further be assumed that bristle number is influenced by only 5 pairs of alleles at 5 different loci. It is certain that genes at more than 5 loci are concerned, but I have chosen a small number to simplify the calculations; in any case, the exact number of loci involved does not greatly alter the conclusions reached. These 5 pairs of alleles can be symbolized by the letters A, a; B, b; C, c; D, d; E, e. Thus B and b represent biochemically different structures occupying a particular place on a chromosome. An individual might then be represented by Aa, BB, cc, Dd, ee, and a gamete produced by such an individual might be $ABcde$, but not $Abcde$, since the individual possesses no b which could be transmitted to a gamete. We will suppose that an individual aa, bb, cc, dd, ee has 20 bristles, and that the replacement of a pair of alleles represented by small letters (e.g. cc) by alleles represented by capital letters (CC) increases the bristle number by 4. Thus individuals aa, BB, cc, dd, ee, or aa, bb, cc, DD, ee would have 24 bristles, AA, BB, CC, DD, EE would have 40 bristles, and so on.

Some assumption must now be made about the number of

bristles possessed by the heterozygote *Aa*, as compared to the homozygotes *aa* and *AA*. It will be shown later that the assumption chosen here makes a big difference to the results to be expected. For the present we shall assume the heterozygote to be intermediate between the two homozygotes; that is, to have two more bristles than *aa* and two less than *AA*. It is also assumed that the effect on bristle number of the alleles at one locus is independent of what alleles are present at other loci.

A further assumption must be made, concerning the mating system. It is often assumed that in a wild population mating is 'at random'; that is, in the present example, the chance of two flies mating together is not affected by the number of their bristles. Now it is known that in human populations this assumption is untrue for some characters. There is a tendency for like to marry like, which is called 'assortative mating'; for example, tall men tend to marry tall women. If this tendency were very strong, the population would gradually divide into two parts, one of tall people and the other of short, between which little or no interbreeding occurred. However, the extent of assortative mating in man is far too slight to lead to such a result, and for some differences, for example of blood group, mating seems to be completely random. It will be necessary to return to the problem of assortative mating in wild populations later, when discussing the origins of breeding isolation between species, but for the present it can safely be ignored.

Finally, some assumption must be made about the relative frequencies in the population of the different alleles, *A*, *a*, etc. If the alleles, *A*, *B*, *C*, *D*, *E* are common and *a*, *b*, *c*, *d*, *e* rare, most individuals in the population will have 40 bristles and individuals with fewer bristles will be rare, increasingly so as the bristle number is reduced. An example of such a situation can be given from the genetics of stature in man. Most people are 'tall' when compared to achondroplastic dwarfs, whose height is about 4 feet, and who occur with a frequency of less than 1 per 10,000 in the population, the condition usually being due to a rare dominant gene. However, for the purpose of our model, it will be convenient to assume that the frequencies of *A* and *a* in the *initial* population are the same, i.e. 50 per cent *A* and 50 per cent *a*, and similarly for the other allele pairs. This corresponds to a situation which is certainly common, in which two alleles at a locus both exist with high frequencies in a population.

With these assumptions we can ask various questions. Later the answers given will be compared with results actually obtained.

(i) With what frequencies will individuals with different numbers of bristles occur in the population? The expected frequencies are shown in Figure 18 (top), which closely resembles that for human stature (Figure 5). Individuals with a bristle number close to the average are common, while extremes are rare. The reason for this is as follows. There is only one genotype which will give an individual with only 20 bristles, namely *aa*, *bb*, *cc*, *dd*, *ee*. There are many different genotypes giving 30 bristles; examples are *Aa*, *Bb*, *Cc*, *Dd*, *Ee* and *AA*, *BB*, *Cc*, *dd*, *ee*. Consequently individuals with 30 bristles are much commoner than those with 20 or 40. The argument is just the same as that concerning the distribution of suits in a bridge hand. No one hand is any more likely to be dealt than any other, but a hand is much more likely to contain 3, 4, or 5 spades than to contain no spades or 13 spades.

It follows that the model will explain satisfactorily the distribution in a population of characters such as size or bristle number.

(ii) How closely will relatives resemble one another? Figure 18 (bottom) shows a typical family obtained by mating two individuals with 32 and 36 bristles respectively. The mean bristle number of this family is 34, halfway between the two parents; to this extent offspring tend to resemble their parents. The variation, or spread, in the family is less than in the population as a whole; this is just another way of saying that sibs tend to resemble one another. The degree of resemblance between relatives can be measured by the 'correlation coefficient': this is a number which would be 1·0 if all the members of a family were identical and zero if there was no tendency for relatives to resemble each other. It can be shown that for this model, the expected correlation between sibs is 0·5, and a similar value of 0·5 is expected for the correlation between parents and offspring.

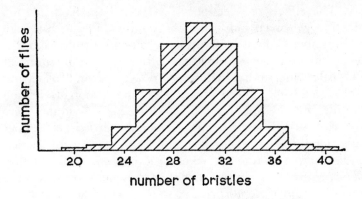

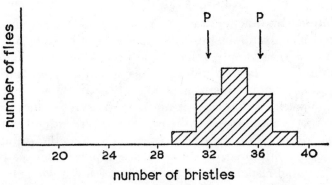

Figure 18. Expected numbers of flies with different numbers of abdominal bristles; (*top*) in a random-mating population; (*bottom*) in a single family, the parents of which had 32 and 36 bristles respectively. These are theoretical distributions deduced from the assumptions outlined in the text.

These expected values agree closely with those obtained for stature in man by Pearson and Lee. Of course this does not prove that the model gives a correct picture of the determination of stature in man. As emphasized earlier, a resemblance between relatives can be partly due to a common standard of life.

So far we have considered only those aspects which can be measured without experimental interference with the breeding system, and for which therefore human data exist. We will now turn to the results to be expected from selection and from inbreeding.

(iii) The effects of selection. What will happen if in each generation we breed only from those individuals with a high bristle number? The most extreme selection possible would be to breed only from those individuals with the maximum number of 40 bristles. Such individuals are homozygotes for all 5 loci and would 'breed true', so that after only one generation of selection we would obtain a population in which all individuals were alike, with 40 bristles. No further increase in bristle number would then be possible. For several reasons, however, selection for quantitative characters never proceeds as rapidly as this. First, even for our simple model, only about one fly in a thousand would have 40 bristles, so that, even in a species in which a single pair can have several hundred offspring, selection as extreme as this would involve a decrease in the size of the population. In practice, far more than 5 loci affect bristle number; if, for example, bristle number were affected by alleles at 15 loci instead of only 5, the proportion of flies in the initial population with the maximum number would be about one in a thousand million. Second, if some of the variation is due to environmental conditions, it is not possible to select flies with the optimal genotype for high bristle number, although it is possible to select flies whose genotype is better than the average, and consequently progress would be less rapid. Other reasons why the initial response to selection may be slower will emerge later.

However, it is possible to select in each generation flies with a bristle number greater than the average for the population from which they were drawn. As shown in Figure 18, the offspring of such flies will tend to resemble their parents. For our model, such selection will lead in a few generations to the same result as the extreme selection described in the last paragraph; we should finish up with a population of flies which were all alike, with 40 bristles. Similarly, by breeding from flies with fewer bristles than the average, we should obtain a population of flies with 20 bristles. The conclusions to be drawn are:

(a) selection would at first lead to a rapid change in the population mean in either direction;

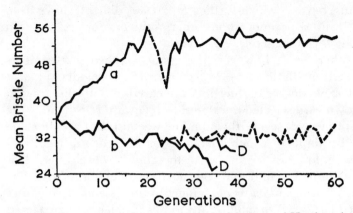

Figure 19. Some results of an experiment by Mather and Harrison, in which, starting from a population of flies with an average of 36 abdominal bristles, lines were selected for increased (a) and decreased (b) bristle number. The dotted lines indicate populations in which selection was relaxed. D indicates populations which died out owing to infertility.

(b) progress under selection would slow down, and finally stop, because there would no longer be any genetic variability for which we could select; and

(c) at this final stage, the population would be much less variable than the initial population.

These conclusions would also hold if the number of loci involved were greater than five, and if there were some environmentally caused variations in bristle number. The effect of either of these factors would be merely to slow down the rate of approach to the final, relatively uniform population. These predictions, based on a very simple model, will now be compared with some actual experiments on selection for bristle number in *Drosophila*. Figure 19 shows some of the results of a classic experiment by Mather and Harrison. Progress was made under selection for bristle number in both directions at first, but finally a limit or 'plateau' was reached, beyond which little further progress could be made. It took considerably longer to reach this plateau than would be predicted from our model, but this is only to be expected for the reasons given above. Similar plateaus, beyond which further progress is difficult, have been reached in experiments on other animals, for example in selecting for increased size in mice.

In other ways, however, the results of these experiments differed from those to be expected from our model. First, a considerable degree of infertility appeared in both the upward and downward selected lines; in the downward selected line the infertility was so severe that the line was lost, but another line derived from it, in which selection was relaxed, survived, and retained a low bristle number in the absence of selection. It is not uncommon to find that continued selection for a particular character brings in its train changes in others and often causes lowered fertility. We shall return to this problem of 'correlated responses' to selection later.

A second difference between experimental results and predictions from the model concerns the variability of the population after selection. It was predicted that the selected population would be much less variable than the initial one. Recent experiments on selection for bristle number by Clayton and Robertson showed that precisely the opposite is the case. They found that at first the variability of the selected populations declined, as would he expected. However, when a level of bristle number had been reached beyond which further progress was difficult, the population was even more variable than the original one. Even more puzzling, they found by comparing close relatives that most of this variation was genetic and not environmental in origin. Clearly, these results cannot be explained by the type of model we have chosen. One or more of the assumptions made must be wrong. However, before deciding which assumptions must be altered so as to fit these new facts, another case will be described in which the experimental results disagree with prediction, since this will provide a clue to the error in our original assumptions.

(iv) The effects of inbreeding. Returning to our model, we can ask what will happen if we select a brother and sister from the population and mate them together, and from their offspring mate together another brother-sister pair, and so on in each generation. We will suppose that a number of brother-sister mated lines are kept, without selecting individuals for bristle number. Consider first a single locus, occupied by the allele *A* or *a*. There are three possible genotypes for an individual, *AA*, *Aa*, and *aa*, and hence there are nine possible types of mating. Of these, two, namely *aa* × *aa* and *AA* × *AA* involve only one of the two alleles. Sooner or later in any brother-sister mated line one of these two kinds of matings will occur; from then on, only one of the two alleles *A* or *a* will exist in the line, and all flies in it will be homozygous for the same allele. The same argument applies to each of the other four loci. Therefore after a number of generations each inbred line will consist entirely of homozygotes for all the loci; a line might consist, for example, only of flies of the genotype *aa*, *BB*, *CC*, *dd*, *EE* or only *AA*, *BB*, *cc*, *dd* *EE*. Which particular allele will be retained at any locus will be a matter of chance, and different lines will be homozygous for different alleles.

The effects of such inbreeding on bristle number will be as follows: all individuals in a given line will have the same number of bristles, so that it will be impossible to alter the bristle number of an inbred line by selection.

Now there are good reasons for believing that in most species inbreeding does in fact lead to genetic homozygosity. For example, as explained on page 134, if skin from one mouse is grafted on to another, a reaction occurs between the graft and the tissues of the host, and the graft is sloughed off. It is known that the differences responsible for the reaction are genetically determined by alleles at a number of different loci. However, if skin from a mouse from an inbred line is grafted on to another mouse of the same strain, no reaction occurs and the graft is retained. This suggests that mice from an inbred line resemble one another in being homozygous for the same alleles at the loci concerned.

Similarly, it is often found that selection is ineffective in an inbred line; this is true of selection for bristle number in *Drosophila*, as would be expected if all the variation in an inbred line is environmental in origin. However, the variation in an inbred line may be considerable, and Rasmuson found that inbred lines are more variable in bristle number than are first generation hybrids between them.[1] It seems therefore that the members of an inbred line may be identical, or almost so, in characters such as those investigated by skin grafting, which are not easily modified by environmental conditions but may nevertheless differ markedly from one another in characters which are more susceptible to such modification. This considerable variability of inbred lines has been demonstrated for size and for rate of development in *Drosophila*, and for many other characters in other species of animals and plants. It contradicts the prediction from our model, that members of an inbred line will be identical, just as did the variability of selected lines observed by Clayton and Robertson. The difference is that in an inbred line there may be little or no genetic variability left, whereas in the selected lines it was found that most of the variation was genetic in origin.

To summarize the argument so far, our simple model will explain the typical pattern of variation of bristle number and the resemblances between relatives, and the fact that a population will change in response to selection for a time but ultimately will cease to respond. It will also explain why it is that inbreeding leads to a genetically uniform population which will not respond to selection. It fails, however, to account for the following observations:

(i) A population which has ceased to respond to selection is often very variable, and much of this variability is genetic. Such populations are often of low fertility and viability.

(ii) Inbred lines can also be very variable, although in this case the variability is not genetic. They are usually of low fertility and viability; an illustration of the effects of inbreeding is given in Figure 20.

How are we to explain these discrepancies? Most of them can be explained if we suppose that some of the genes influencing

---

[1] The first generation hybrids between two inbred lines receive a complete set of chromosomes from each parental line and therefore resemble one another genetically as closely as do the members of an inbred line. The comparison made by Rasmuson is therefore a fair one. Second generation hybrids may receive genes from the two parental lines in varying proportions, and are usually more variable in phenotype than the first generation hybrids.

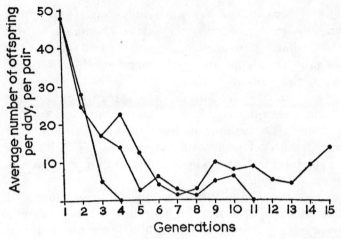

Figure 20. The productivity of three brother-sister mated lines of *Drosophila subobscura*, descended from a single wild-caught female.

bristle number also alter fitness (viability, fertility etc.) in the following way:

genotype	*aa*	*Aa*	*AA*
Bristle number	low	medium	high
fitness	medium	high	low

In other words, the genotype with the maximum bristle number would have a low fitness. If there were a number of such genes, selection for high bristle number would cease to make progress because of the opposing effects of natural selection, the optimal genotypes from the point of view of bristle number having such a low fitness that they failed to survive. The population would continue to be genetically variable after it had ceased to respond to selection.

The same assumption will account for the effects of inbreeding shown in Figure 20. Inbreeding necessarily leads to an increase in the number of loci at which individuals are homozygous, and hence to a lower fitness. Since different inbred lines will, by chance, become homozygous for different alleles, the F_1 hybrids will tend to be heterozygotes; they are usually of high fitness, which is what we would predict. The only observation not explained by this modification to our assumptions is that inbred lines tend to be very variable in phenotype. The reason for this appears to be that heterozygous organisms are well stabilized or buffered against accidents during development, whereas inbred ones are easily thrown off course.

We have been led to assume that for at least some loci the heterozygote is fitter than either homozygote; such loci are said to be 'heterotic'. The assumption has enabled us to account for a number of features of the experimental results. We shall return in the next chapter to discuss why it should be that heterotic loci should be so common.

In the original model for the inheritance of bristle number I assumed what is called 'additive' inheritance; that is, I assumed that heterozygotes are intermediate in bristle number between homozygotes and that the effects of genes at different loci can be added together. These assumptions work rather well for bristle number, and discrepancies arise only after prolonged selection. But if the selected character is itself an important component of fitness, the additive assumption does not work at all. Consider for example the results of experiments on development rate (measured by the number of days taken to develop from egg to adult) in *Drosophila subobscura*. Selection for increased development rate is almost completely ineffective. Inbreeding always leads to slow development; F_1 hybrids between inbred lines develop much faster than either parent. These facts only make sense if one supposes that genetic homozygotes develop slowly and heterozygotes fast.

Haldane has pointed out that there is a good reason why most of the genetic variability of characters which contribute to fitness should be due to genes with heterotic effects, whereas the variability of characters such as bristle number, which are trivial as far as fitness is concerned, should be due in the main to genes with additive effects. Wild populations are the result of a long period of natural selection for greater fitness. They have therefore already reached a 'plateau' as far as general fitness is concerned, and their genetic variability resembles the residual variability in a line selected for increased bristle number after such a plateau has been reached. Thus if a pair of alleles A and *a* have additive effects on fitness, so that, let us say, *AA* is fitter than *Aa*, which is in turn fitter than *aa*, the allele *a* would be reduced to a very low frequency in the population by natural selection and so would contribute little to the variability of the population. If the heterozygote is fitter than either homozygote, both alleles will be common in the population, and will contribute appreciably to its variability.

The situation is very different as far as bristle number is concerned. It is quite possible that a pair of alleles, *B* and *b*, might have additive effects on bristle number but heterotic effects on fitness, and in this case both alleles would be common in wild populations. If many such pairs of alleles were present, artificial selection would be able to produce rapid changes in bristle number in either direction but only at the expense of a lowering of the fitness of the population by making it genetically more homozygous.

There is one final complication which must be mentioned. Genes do not always assort independently. Their loci may be situated on the same chromosome; that is, they may be linked. What are the effects of this phenomenon on variation and selection? Still sticking to the example of bristle number, suppose that alleles *A* and *B* increase, and *a* and *b* decrease bristle number. Now, if selection is started in a relatively small population, it is quite possible that only chromosomes of the types *Ab* and *aB* will be present in it. If so, there are only three possible genotypes, namely *Ab/Ab*, *aB/aB*, and *Ab/aB*. All three genotypes produce individuals with the same number of bristles. Therefore these two loci contribute nothing to the variability of bristle number in the population, and selection cannot utilize them. But sooner or later crossovers between the two loci will occur, giving rise to chromosomes *AB*, *ab*, and it is then possible to select individuals *ab/ab* with fewer bristles, or *AB/AB* with more bristles than were possessed by the original population.

Thus there was in the original population some concealed but potential variability, which could only become apparent and available for selection as a result of crossing over between linked loci. Now on page 157 it was described how, in a population selected for increased bristle number, Mather and Harrison observed at first rapid progress under selection, but soon obtained a population in which further selection was ineffective. One such population, however, which had reached such a plateau and which had not changed under selection for some sixty generations, then showed a further slight but real increase in bristle number before again coming to a stop. They explain this by suggesting that one or more crossovers took place, making further genetic variability available for selection to act upon.

It is difficult to decide how important the effects of linkage may be for the genetics of populations. For example, earlier in this chapter I suggested that at many loci the heterozygote is fitter than either homozygote. The facts could be explained equally well if the heterosis is between blocks of tightly linked genes rather than between alleles at a single locus. We shall meet this difficulty again on page 191.

The most important general conclusion for evolution theory to be drawn from experiments on artificial selection is that it does work. For almost every character that has been studied in an outbreeding organism, artificial selection has produced large and rapid changes in the characteristics of the population. This is an indication of the extensive genetic variability of outbreeding populations; it is this variability which is the raw material of evolution. There is, however, one respect in which artificial selection experiments are a poor model of evolution. In experimental populations, changes occur at a rate many orders of magnitude greater than the rates characteristic of evolution (see page 277), but they continue only for a time, and cease when all the available genetic variability has been fixed. No further progress can be made until new variability has been generated by mutation. In evolution there must in the long run be a balance between the fixation of genetic variation by selection and the generation of new variation by mutation.

CHAPTER 10

Natural Selection in Wild Populations

We should expect to find the most rapid evolutionary changes in populations suddenly exposed to new conditions. It is therefore natural that one of the most striking changes which has been observed in a wild population has occurred in the industrial regions of England and western Europe during the past hundred years. This is the phenomenon of 'industrial melanism', the appearance and spread of dark forms of a number of species of moths.

Melanic forms were first reported in 1850 from Manchester in the Peppered Moth, *Biston betularia*. Today the original form of this moth, coloured for concealment, is rare in the industrial north, and dark forms are common in many country districts also. A similar replacement of speckled by dark-coloured moths has since occurred in industrial areas in a number of other species. These species are not particularly closely related, belonging to five different families, but they have in common the habit of resting on the bark of trees, and a speckled grey and brown colouring resembling that of lichens or of bark.

In all cases the dark colour is dominant over the cryptic and in most the difference is due to a single dominant gene. Thus in *Biston betularia* the dark variety, *carbonaria*, which is a uniform black with a pair of white spots at the base of the fore-wings, differs from the speckled form by the presence of a dominant gene, and the same is true of a less common dark variety, *insularia*, of the same species.

Now in industrial areas the bark of trees, and indeed most other objects upon which a moth might settle, become begrimed with soot. It may be that in such conditions the dark forms are

165

less easily seen than the pale, whereas in country districts the latter are difficult to detect when settled on the bark of a tree, against which the black forms stand out vividly. If this were so, then selective killing by insect-eating birds could explain the spread of melanic forms in industrial areas. Kettlewell has been able to show that selection can in fact work in this way. His experiment is worth describing, since it illustrates one of the ways in which the occurrence of natural selection can be demonstrated.

He released a number of Peppered Moths of the normal and *carbonaria* varieties in a bird preserve near Birmingham which is polluted by soot. It was first necessary to satisfy himself that the moths were in fact being eaten by birds. This was done by watching individual moths through binoculars when they had settled after release, and observing that they were being taken by robins and by hedge sparrows. However, it was not easy to discover by these means which form was found most easily by the birds. Therefore each moth released was marked by a small spot of paint, and a number of moths were subsequently recaptured with a light trap at night. The numbers released, and subsequently recaptured, were as follows:

	Released	Recaptured
Carbonaria	416	119
Pale	168	22
Total	584	141
% *Carbonaria*	71	84
% Pale	29	16

Thus of 584 moths released, 141 were recaptured; of course, not all the moths which were released but not recaptured had been eaten by birds. Many were doubtless still alive but did not come to the trap. Consequently, it is impossible to say what proportions of the two forms were taken by birds. However, it is reasonable to assume that those recaptured were a fair sample of those left alive, and hence that the proportion of pale forms had fallen in the interval between release and recapture from 29 per cent to only 16 per cent. Since observation showed that the moths were in fact being taken by birds, it is fair presumption that the change was due to selective killing by birds.

Thus in a soot-polluted area it has been demonstrated that the melanic form of the Peppered Moth is at a selective advantage because it is less easily seen by birds. It has similarly been shown that the reverse is the case in unpolluted areas. More recently, Kettlewell has been able to show experimentally, using black and white squares, that the carbonaria variety tends to settle on a dark background, and the pale variety on a light background. Thus the advantages of protective coloration are enhanced by appropriate behaviour.

These experiments would explain the spread of melanic forms in industrial areas, but do not account for the increased numbers of melanic forms of the Peppered Moth, and of certain other species, in some unpolluted districts. This, however, may be explained by some experiments by Ford on the Mottled Beauty, *Cleora repandata*, of which a black variety, due to a single Mendelian dominant, has spread in northern industrial areas. Let us denote the allele producing the dark colour B, and the normal allele b. Then b/b moths are speckled in colour, and B/b and B/B moths black. Ford counted the offspring from the cross $B/b \times b/b$, from which the Mendelian expectation is a ratio of one black to one speckled. Small departures from this ratio are to be expected by chance, but larger departures would suggest that one or other form has a lower viability. When the caterpillars from such families were well fed, there was no significant departure from a 1:1 ratio, but when they were starved on alternate days, the numbers were 51 melanics and 31 speckled. These figures can be explained if, although zygotes were formed in the ratio 1 B/b : 1 b/b, a greater proportion of the b/b caterpillars died before becoming adult.

This result suggests that the melanics have a physiological advantage over normal moths, at least in conditions of food shortage, and that they are only prevented from increasing in frequency by the contrary selection due to birds, which find the melanic forms more conspicuous in country districts. Where, as in industrial areas, the direction of the latter type of selection is reversed, the melanics increase in frequency. However, in country districts the two types of selection may be nicely balanced, although usually favouring the speckled forms. With the changes in conditions in the countryside resulting from man's activities, it may be that this balance has been shifted in some areas in favour of the melanics; this is the more likely because soot pollution may extend great distances from the centres of industry.

One other problem raised by industrial melanism is worth discussing further: why is it that in all cases where melanics have established themselves as the most frequent variety in any area, they have proved to be due to dominant genes? In a number of species melanic varieties due to recessive genes are known, but in no case have they become common in nature. The explanation is as follows. The genes responsible for melanism, whether dominant or recessive, were rare before the development of industrial areas. Now if a recessive gene is rare in a population, only a small proportion of the few recessive genes present are exposed to selection, since most of them occur in heterozygous condition with their dominant alleles, and so produce no phenotypic effect on which selection could act. In the case of a rare dominant gene, all the genes in the population produce a phenotypic effect and so are exposed to selection. Thus an increase in the frequency of an initially rare recessive gene will at first be very slow, even if it confers a large selective advantage in homozygous condition, whereas the increase in frequency of an initially rare but advantageous dominant gene will be quite rapid. The spread of industrial melanism has been very rapid, and it is therefore not surprising to find that it is due to dominant genes. It does not follow, of course, that rare recessive genes can never spread through a population, but only that if they do spread, they will do so more slowly.

There is, however, another possibility in the case of recessive genes which are also advantageous. If such a recessive gene is rare, then there will be selection in favour of genetic changes at other loci which make the advantageous gene effective in heterozygotes; that is, which change the initially recessive gene into a dominant. This idea of the 'evolution of dominance' was due to Fisher, and it may explain why alleles which are common in wild populations are more often than not dominant over rare mutant alleles. Selection has acted so as to render the more favourable alleles dominant and the less favourable ones recessive.

This contrast between the effects of selection on a rare dominant and on a rare recessive gene depends on the fact that there is a difference between the frequency of genes in a population and the frequency of the corresponding genotypes and phenotypes, as the following argument will make clear. A very simple mathematical relationship between the frequency of genes and of genotypes was discovered independently by Hardy and by Weinberg, and is often referred to as the Hardy-Weinberg ratio. This law is so important in arguments about selection in wild populations that, at the risk of irritating readers who dislike even the simplest algebra, it will be explained here.

The law holds only if mating is at random, a situation which is almost or quite true in most cases. Suppose that there are two alleles, A and a, at a particular locus, and that their frequencies in a population are p and q respectively, where $p + q = 1$. If, for example, A were nine times as common in the population as a, then p would be 0·9 and q 0·1. The probability that an individual receives the allele A from his father is then p. If mating is random, there is a similar chance p that he also receives an allele A from his mother. Hence the chance of an individual receiving A from both parents is $p \times p = p^2$, which is therefore the proportion of A/A individuals in the population. By an exactly similar argument, the proportion of a/a individuals is q^2, and of A/a (or a/A) individuals is $2pq$. Some typical numerical values, assuming that A is dominant to a, are as shown in the table on the next page.

Thus if a recessive gene a has a frequency of 0·01, or 1 in 100, then only one individual in 10,000 will be homozygous for it, and so be exposed to selection. In the case of a dominant gene A with a frequency of 1 in 100, one individual in 50 (approximately) will show the effects of the gene. This makes more precise the argument above concerning the origin of

	Gene Frequencies		Genotype Frequencies		
	A p	a q	A/A p^2	A/a $2pq$	a/a q^2
Rare recessive	0·99	0·01	0·9801	0·0198	0·0001
Frequency of dominant phenotype				0·9999	
Rare dominant	0·1	0·99	0·0001	0·0198	0·9810
Frequency of dominant phenotype				0·0199	
Equally frequent alleles	0·5	0·5	0·25	0·5	0·25
Frequency of dominant phenotype				0·75	

industrial melanism through dominant mutations. Another important application of the Hardy-Weinberg ratio will be described later in this chapter.

If there has been time for a striking change to occur in a number of species of moths in the last hundred years, why is it that the characteristics of most wild populations have changed so little during the same period? The answer to this question is at least in part that the constancy of wild populations is more apparent than real. Moths have attracted the attention of many amateur naturalists, so that changes are more likely to be recorded in them than in most other wild populations. Also, a change in coloration would be noticed, whereas a change in physiology or even in form might be missed.

Other rapid evolutionary changes have, however, been observed. In all cases they concern adaptations to changes in the environment produced by man; among the most dramatic are the evolution of insecticide resistance in insects and of drug resistance in bacteria. All the same, the fossil record shows that the rates of evolutionary changes have been very slow compared to the rates of change in domestic animals or in laboratory selection experiments. Some of the reasons for the slowness of evolutionary changes will now be described. It will become apparent that the constancy of wild populations is not to be accounted for by the absence of natural selection, but rather that natural selection is as effective in maintaining the constancy of a population as it is in changing it.

It was the crux of Darwin's argument that selection adapts a population to its conditions of life. It is a necessary consequence, at least for many characteristics, that the typical or average members of a population are the best adapted, and that extremes in any direction are at a disadvantage. For example, in Chapter 2 evidence was given to show that it is a disadvantage for a swift to lay too many eggs, just as it is a disadvantage to lay too few. A classic illustration of the action of selection against the more extreme members of a population is due to Bumpus, who measured the wing length and other bodily dimensions of sparrows killed in a storm. Compared to the survivors, the sparrows which died were more variable, including an excess of birds with wings markedly shorter or longer than the average. Similarly, Rendel found that duck eggs which failed to hatch had almost the same average size, but were far more variable than eggs which did hatch. The same principle has been demonstrated for the birth-weight and survival of human babies by Karn and Penrose, and for the shapes of snail shells by Weldon.

Now if it is usually the extreme forms of a population which are killed off, one might expect that the population would become less and less variable, until in time all individuals were genetically identical, and varied only from the influence of environmental conditions. Undoubtedly selection does limit the range of variation in a wild population, but there is no reason to believe that it results in a continuous decline in variability, and plenty of reasons to believe that it does not; whenever a wild population has been studied, an abundance of genetic variability has been uncovered. How is it, then, that selection against extremes has not led to genetic uniformity? Three reasons for this will be discussed:

(a) Different types may be favoured by selection at different times or in different places.
(b) The fittest type may be a heterozygote.
(c) The fitness of genotypes may vary with their frequency, increasing as they become rarer.

An example of the first is afforded by colour polymorphism in the land snail *Cepaea nemoralis*, common in open country and in varied types of woodland in England, and in fact over much of Europe. The shell may be yellow, pink, or brown, and is also characterized by the presence or absence of a series of black bands. These differences are known to be determined genetically, and not by the influence of the environment. Diver has shown that the banding polymorphism has existed at least since neolithic times. The polymorphism is therefore stable and long-lasting, and not a temporary transition stage as in the case of industrial melanism. Our knowledge of the selective forces which have maintained this stability is mainly due to the work of Cain and Sheppard, whose findings are summarized below.

The frequencies of the different kinds of shells vary according to the habitat in which the snails are living. Where the background is comparatively uniform, the proportion of un-banded shells is high. Thus in short turf in the open, and also in beech woods where there is little undergrowth and a uniform floor of dead leaves, the proportion of unbanded shells is usually greater than 90 per cent, whereas in hedgerows and rough herbage it is usually less than 40 per cent and often as low as 10 per cent. Intermediate frequencies of banded and unbanded shells are found in oak woods and mixed deciduous woodland where there is much undergrowth. The frequency of shells with a yellow ground colour rises with the greenness of the background, being higher in open country than in thick woodland with little undergrowth and a predominant back-ground of bark and dead leaves.

This association between colour and banding of the shell and the nature of the background suggests that the variations may afford protection against some animal hunting by sight. Such is the Song Thrush, which has the habit, convenient for the investigator, of breaking the shells of captured snails on particular stones, or 'anvils'. By examining the shells at these anvils, Cain and Sheppard have been able to compare the frequencies of different kinds of snails being killed by thrushes with the frequencies in the population in the same area.

The fraction of yellow shells at thrushes' anvils was recorded in a mixed woodland from April to July. In April this fraction was higher than the fraction of yellow shells among snails collected in that wood, whereas in July the fraction of yellow shells among snails killed by thrushes was lower than for the population as a whole. Thus early in the year selection was favouring snails with pink and brown shells, whereas later, when the background was greener, selection was favouring snails with yellow shells.

Sheppard has similarly been able to demonstrate selection in favour of unbanded shells on a uniform background. He obtained the following figures for living snails, and for snails killed by thrushes:

| | Number of shells | | | |
	Banded	Unbanded	Total	% Banded
Living	264	296	560	47·1
Killed by thrushes	486	377	863	56·3

Thus although slightly less than half the snails in this population were banded, more than half the snails killed by thrushes were banded. If this were the only selective force operating, it would suffice to reduce the banded forms to a rare variety in a relatively short time.

We have now to explain how it is that the polymorphism for shell colour, although influenced by selective killing by thrushes, has nevertheless been preserved for thousands of years. There are several reasons. First, even in a single locality selection does not always favour the same type; in mixed woodlands the pink and brown shells may be at an advantage early in the year but at a disadvantage later. Such a balance of selection could not by itself produce a stable and long-lasting polymorphism, but it would help to delay the disappearance of the type which was on the average the less fit.

However, in other habitats the same coloured shell should be favoured throughout the year; for example, we would expect yellow unbanded shells to be the best protected at all seasons on short turf, and pink or brown unbanded shells in beech woods. Yet, although these types are the commonest in their respective habitats, other types of individual also occur there. This is in part due to migration and to interbreeding between snails from different habitats. This, however, is not a sufficient explanation. Cain and Sheppard studied a locality where a beechwood bordered open country with short grass. They found a typically low-frequency of yellow shells in the wood, and a typically high frequency on the grass. There was a zone of interbreeding along the border of the wood in which intermediate frequencies of yellow shells occurred, but this zone was only some fifty yards wide. It seems, therefore, that the range of movement of individual snails is too small to explain the occurrence of colour polymorphism in more isolated colonies, and therefore some other explanation must be sought for the presence in such colonies of snails whose colouring renders them more conspicuous.

But before seeking such an explanation it must be emphasized that in many other animal species, in which the range of movement of individuals is greater, such movement, accompanied by interbreeding, may render it impossible for a species to become divided up into a series of local populations each adapted to a particular habitat. In such cases natural selection favouring different genotypes in different localities will result, not in a series of genetically different, locally adapted populations, but in a single interbreeding population with considerable genetic variability between individuals. Movement and interbreeding have a levelling-out effect; a restricted range of movement favours the formation of differentiated, locally adapted populations. In either case, however, selection in favour of different types in different places tends to maintain the genetic variability of the species as a whole.

We must now return to the problem of the maintenance of polymorphism in local populations of *Cepaea*. It was pointed out in Chapter 9 that selection for a particular type will not lead to genetic uniformity if the selected type is a genetic heterozygote, since such heterozygotes do not breed true. Now there is some evidence that the genes responsible for the colour and banding of the shell in *Cepaea* also have physiological effects, and that the most vigorous individuals are genetic heterozygotes for these genes. This fact probably accounts for the occurrence of snails with yellow or with banded shells in beech woods, or with brown shells on grass. However, the superior fitness of hetero-zygous genotypes is of very general occurrence, and some examples will be given where the evidence is clearer than in the case of *Cepaea*.

The fact that inbreeding usually leads to a decline in vigour, which can be restored by outcrossing, can most easily be explained on the assumption that genetic homozygotes, for at least some loci, are inferior in fitness to heterozygotes. One of the clearest examples of this phenomenon for a particular locus in a wild population is known in man. In certain parts of Africa the disease 'sickle-cell anaemia' is common. The red blood cells of sufferers from this disease are distorted, often sickle-shaped, and the condition is usually fatal early in life. Affected individuals are known to be homozygous for a gene which we will call s, its normal allele being called S. The heterozygote Ss can be recognized from a slight sickling of their red cells when the blood is deoxygenated, but they do not suffer from anaemia. The frequency of heterozygous Ss individuals in some areas rises as high as 40 per cent, and it had long been a puzzle how such a high frequency of the allele s could be maintained in face of the intense selection against the ss homozygotes. The answer has been found by Allison, who has shown that the frequency of the S allele is particularly high in those parts of Africa where tertian malignant malaria is prevalent, and that heterozygous Ss individuals are much more resistant to this disease. Thus the heterozygous individuals are fitter than ss homozygotes because they do not suffer from anaemia and fitter than SS homozygotes because they do not suffer from malaria. Selection against s in homozygotes is just counter-balanced by selection in favour of s in heterozygotes so that s is maintained at a fairly high frequency in spite of its lethality in the homozygous condition.

This example is an extreme one in that one of the possible homozygotes is lethal. However, a similar equilibrium will be reached between any two alleles, A and a, provided that the heterozygote *Aa* is fitter than either homozygote. This is an example of heterosis; the use of capital and small letters does not here imply dominance, since all three possible genotypes are different in their phenotypes. That the equilibrium is stable can be seen from the following argument. If *a* is rare in the population, it will usually occur in heterozygous *Aa* individuals, which are fitter than the average, and so more *a* genes will be transmitted to the next generation; thus when *a* is rare it will tend to increase in frequency. Similarly, *A* when rare will tend to become more frequent. There will be some intermediate frequency of *A* and *a* at which there is a stable equilibrium.

Heterosis may well be the commonest cause of genetic variability in outbreeding populations. This does not mean that most new mutations have heterotic effects in combination with existing alleles; in fact Stern found only one or two such cases out of seventy-five studied. It does mean that new mutations with heterotic effects are likely to survive and to be established with high frequencies in the population, whereas other muta- tions will either remain rare, or, less often, will replace existing alleles which will in turn become rare.

A method will now be described for detecting the presence of alleles with heterotic effects in wild populations. If mating is at random, zygotes *aa*, *Aa*, and *AA* will be formed with frequencies corresponding to the Hardy-Weinberg ratio (see page 169). If, however, the alleles have heterotic effects, a greater proportion of *Aa* individuals will survive to become adults. It is therefore a feature of heterosis that in an adult population heterozygotes should be present in greater numbers than would be expected from the formula. Such an excess of heterozygotes can be accepted as a demonstration of heterosis, provided that there is evidence that mating is at random.

This and other techniques have been applied by Dobzhansky and his colleagues in the study of selection in populations of *Drosophila pseudoobscura*. The examples of natural selection described earlier in this chapter were concerned with colour differences which afford protection against a predator whose eyesight is similar to our own. The examples were chosen because they have been studied in some detail, mainly because the differences are easy to see. In Dobzhansky's work, however, the genetic differences concerned cannot be recognized from the structure of adult flies, but concern instead differences in the structure of the chromosomes, which can only be recognized in the larvae.

In Chapter 7, it was explained that if two breaks occur in a chromosome. the central piece may be rejoined to the end pieces in an inverted position, so that an initial order, say 1234567, is converted to 1254367. Such inversions probably occur as rare events in all animal and plant species, but rarely become established because individuals with two chromosomes of different orders tend to have a reduced fertility (see page 264). However, in the two-winged flies (*Diptera*), no such lowered fertility occurs, and inversions have a correspondingly better chance of becoming established.

In *Drosophila*, the presence of two chromosomes with different gene orders can be detected in the larvae by examining the giant chromosomes of the salivary glands. In these glands, pairs of homologous chromosomes come to lie side by side, appearing as a single thread, each element of one chromosome being opposed to the corresponding element of the other. If two chromosomes have different gene orders, they can only pair in this way by forming an 'inversion loop', as is shown in Figure 15.

In a population in which two chromosome orders exist, let us say the 'standard' or *ST* and 'arrowhead' or *AR* orders, three types of larvae can be distinguished, namely the 'structural heterozygote', *ST/AR*, in which an inversion loop is formed and can readily be seen, and two kinds of 'structural homozygotes', *ST/ST* and *AR/AR*, which can be recognized, though with greater difficulty, by examining the banding patterns on the giant chromosomes, which are arranged in a different order in the two types. The chromosomal structure of an adult fly, however, can only be discovered by mating that fly to another of known structural type, and then examining a number of the larval offspring.

Now the origin of any given inversion is a single event, occurring in a single chromosome of an individual animal, the inverted chromosome subsequently being reproduced in the changed form. Such an inversion will include the loci of many hundreds and possibly thousands of genes. If an inversion at its first occurrence happens to include a number of favourable alleles, it may become established in the population. Now it was shown by Sturtevant that there is little exchange of genetic material (i.e. little genetic crossing over) between chromosomes with different gene orders. Therefore the inverted chromosomes in a population will tend to carry a different constellation of alleles to those carried by the uninverted chromosomes, partly because the original inverted chromosome will have included rare alleles in at least a few loci, and partly because of new mutation. Consequently structural heterozygotes, say *ST/AR*, will also be genetically heterozygous for alleles at a number of loci for which *ST/ST* and *AR/AR* individuals are homozygous. The value of these inversions to the investigator, therefore, is that they make it possible to recognize genetic differences, not at a single locus, but at many loci at once, even although the effects of the different alleles at these loci cannot be recognized by examining the structure of adult flies.

In the American species *D. pseudoobscura* the third chromosome exists in a number of different orders in wild populations, and consequently structural heterozygotes are common. The causes maintaining this polymorphism for chromosome order have been studied in 'population cages', in which a large population is maintained in very overcrowded conditions for many generations. In one such experiment, the initial population contained flies with two different types of chromosomes, Standard, *ST* and Chiricahua, *CH*, in the proportions 11 per cent *ST* to 89 per cent *CH*. Samples of larvae were taken from the cage at intervals and examined, and the percentage of *ST* chromosomes present was estimated. The results are shown in Figure 21. At first the frequency of *ST* chromosomes rose, suggesting that *ST/ST* individuals were fitter than *CH/CH*. After about eight months, equivalent to rather more than eight generations, the frequency of *ST* chromosomes reached about 70 per cent, after which it showed no further increase, suggesting that a stable frequency had been reached.

Such stability could be explained if *ST/CH* individuals are

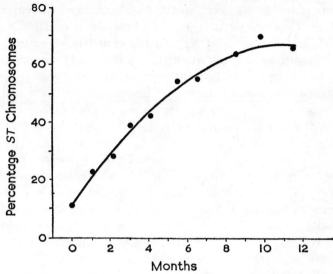

Figure 21. The percentage of *ST* chromosomes in an experimental population of *Drosophila pseudobscura* (after Dobzhansky).

fitter than either structural homozygote. This was confirmed by taking adult flies from the cage and determining their structural type. It was found that structural heterozygotes occurred in excess of the frequency to be expected from the Hardy-Weinberg ratio. Such an excess could be due to the fact that mating was not at random, homozygotes tending to mate with individuals unlike themselves. However, this explanation was ruled out by taking eggs from the population cage and raising the larvae which hatched under optimal conditions, so that almost all survived to become adults. The frequencies of the different types among these adults were found to fit the Hardy-Weinberg ratio. The only remaining explanation of these facts is that, at least in population cages, a stable polymorphism is maintained by the superior fitness of the heterozygotes. It has been possible to show that there is a similar excess of structural heterozygotes among males caught in the wild; unfortunately a similar demonstration is not possible for wild females, because there is no simple way of determining the structural type of a female which has been fertilized before capture.

The population cage experiment described above was carried out at 25 °C. The fact that an equilibrium was reached with frequencies of 70 per cent *ST*: 30 per cent *CH* suggests that *ST/ST* homozygotes are fitter than *CH/CH*. However, this superiority is apparent only at high temperatures; at 16 °C. there was no change in the initial frequency, so that a stable equilibrium existed with a greater proportion of *CH* than of *ST* chromosomes. In wild populations it is found that the frequency of *CH* increases at the expense of *ST* from March to June, followed by an increase in the frequency of *ST* during the hot weather from June to August. The relative frequencies of the two chromosome orders also vary with height above sea level, *ST* being commoner at low levels and decreasing in frequency with height above sea level. Therefore, as in the case of colour varieties of *Cepaea*, selection is favouring different types in different circumstances, but here the differences are of climate rather than of background.

Chromosomal polymorphism is not confined to *D. pseudoobscura*, but occurs in the majority of *Drosophila* species, sometimes in a more extreme form. In a population of *D. tropicalis* studied by Dobzhansky and Pavlovsky, there are two common chromosome orders, and both types of structural homozygote die early in development, only the heterozygotes surviving to breed. Although this means that half the zygotes formed are inviable, the population is a flourishing one. In the European species *D. subobscura*, chromosome polymorphism is not confined to a single chromosome pair, as it is in *pseudoobscura*. Inversions are common on all the long chromosomes, and it is rare to find an individual structurally homozygous for all chromosomes. Selection against homozygotes is so intense that even prolonged inbreeding does not always lead to structural homozygosity.

Inversions are uncommon in most groups, because their presence reduces fertility by disorganizing meiosis. They are common in *Drosophila* and other two-winged flies because these possess special mechanisms. which prevent loss of fertility. They have been important in population genetics in telling us something about the effect, of linked blocks of genes, but they can tell us little about the effects of genes at particular loci. Thus let us call the alleles on one chromosome order *ABCD*, and on the other *A'B'C'D'*. If the heterozygote is fitter than either homozygote, it may be because of heterosis at individual loci – that is, because *AA'* is fitter than *AA* or *A'A'*, and so on. But one could equally well have heterosis for the block of genes as a whole but not for any particular locus. Until recently there were rather few cases in which we could study genetic variation at individual loci in natural populations (blood group loci provided one such case). With the advent of protein electrophoresis, discussed in the next chapter, this difficulty has largely disappeared.

I now turn to the third of the processes listed on page 171 which can maintain genetic variability; this is the process of frequency-dependent selection, whereby the fitness of a genotype increases as its frequency decreases. There are a number of reasons why this may be so; I shall discuss examples arising from predation, from diseases, and from competition for limited resources.

It has been shown that a predator hunting by sight may form a 'search image' of its prey, and hence find only prey items resembling that image, ignoring other equally edible items. Clarke has shown that wild birds form such search images, and that an individual bird is more likely to form an image of a common than a rare item. This is only what one would expect, since the initial formation and subsequent retention of a search image depends on being able to find the corresponding food. In the case of a polymorphic prey species such as *Cepaea*, rare forms will be at an advantage because thrushes and other birds will tend to find only the common forms. Clarke has argued that this may be the explanation for the presence of apparently conspicuous forms, such as snails with yellow shells in beech woods.

Haldane pointed out that variability would be maintained if disease-causing organisms were adapted by natural selection to attack the common biochemical types in a population but not the rare variants. This kind of thing is known to happen in wheat, in which rusts adapt themselves to attack the varieties most commonly grown, so that periodically it becomes necessary to seek new strains of wheat resistant to the prevalent strains of rust and by crossing, to confer their resistance on the commoner high-yielding varieties.

A third cause of frequency-dependent selection arises if there is variability within a species in ability to make use of limiting resources. To fix ideas, suppose an insect lays eggs on two food plants, and that there are genotypes adapted to the different plants. Levene showed that genetic polymorphism can be maintained in such a situation, even if individuals raised on the two plants mate at random. This is a form of frequency-dependent selection for the following reason. If the genotype adapted to a particular plant is rare, it suffers less competition than if it is common, and hence its chances of survival are better. The same argument applies whenever two genotypes are adapted to utilize two different limiting resources.

This kind of situation has been widely invoked to explain genetic polymorphism. Indeed, whenever someone argues that a species will be genetically more polymorphic in a varied environment than a uniform one, this is the mechanism they have in mind. It is important to distinguish between cases in which the environmental 'patches' are large relative to the range of an individual ('coarse-grained' in Levins's terminology) so that an individual raised in one patch will usually breed in the same patch, and cases in which the patches are small ('fine-grained' in Levins's terminology), so that individuals raised in one patch will breed in another. The former cases fall under the first heading on page 171; no difficulty arises in explaining genetic variability, but we would expect to be able to detect different gene frequencies in different patches, as can often be done for shell colour in *Cepaea*. The latter, fine-grained, cases must be explained by frequency-dependent selection of the type now being considered. I have pointed out two reasons why one must be cautious in suggesting this explanation. The first is that the resources must actually limit the population, otherwise no frequency-dependent effect arises. The second is that it can be shown (but only by the use of algebra which would be out of place here) that the selective advantages needed to maintain a polymorphism are large – certainly 20 per cent and perhaps as high as 50 per cent. For these reasons I doubt whether in outbreeding species many loci are maintained in a polymorphic state by this process. Perhaps its main interest is that such a polymorphism could be the starting point for speciation without geographical isolation; I shall return to this point on pages 239–44.

CHAPTER 11

Protein Polymorphism

A major difficulty in population genetics is that our theory has to do with the frequencies of genes and genotypes in populations whereas our observations are of phenotypes. Only rather rarely do we know the genetic basis of the phenotypic differences we observe. Even when we know that a phenotypic difference is caused by alleles at a single locus, the phenomenon of dominance usually prevents us from identifying genotypes. If a gene is represented by only one allele in a population, we have no way of knowing that it exists at all, because we recognize genes by the differences they cause.

In principle, molecular biology has provided a way out of these difficulties. If we knew the amino acid sequences of a particular enzyme in all the individuals of a population, we would be able to identify all the relevant alleles in the population and their frequencies. We could even recognize the presence of a gene which did not vary. Unfortunately the labour involved in such an enterprise makes it quite impractical. There is, however, a relatively cheap and simple procedure – gel electrophoresis – which enables us to recognize at least some of the differences between proteins. A crude and unpurified extract of an individual organism is placed on a gel in a suitable solution and an electric field is applied. Different proteins move through the gel at different speeds, depending on their size and configuration and on their electric charge. Thus all the soluble proteins are spaced out linearly over the gel. The gel is then 'stained' for the presence of a particular enzyme by providing the enzyme with a suitable substrate and arranging that a colour change is associated with the enzyme-catalysed reaction. In this way,

184

Genetic Variability in a Number of Organisms

Species	Number of Loci	Proportion of Loci Polymorphic per Population	Proportion of Loci Heterozygous per Individual
Homo sapiens	71	0·28	0·067
Mus musculus (house mouse)	41	0·29	0·091
Peromyscus polionotus (American Deermouse)	32	0·23	0·057
Drosophila pseudoobscura	24	0·43	0·128
6 other *Drosophila* spp.	18–31	0·25–0·86	0·076–0·184
Limulus polyphemus (Horseshoe Crab)	25	0·25	0·061
Silene maritima (Bladder Campion)	21	0·29	0·149

coloured bands appear on the gel at points where enzymes with the corresponding catalytic activity were present.

In general, the technique can only be used to distinguish between proteins which differ in electric charge. Of the twenty amino acids in proteins, two (arginine and lysine) are positively charged, and two (aspartic and glutamic acid) negatively charged, the remaining sixteen being uncharged. Only those substitutions which replace an uncharged amino acid by a charged one or vice versa will be recognized. It can be calculated that between one third and one quarter of all substitutions will be recognized.

The first systematic attempts to use this technique to study the degree of variability in natural populations were by Harris on human populations and by Lewontin and Hubby on *Drosophila pseudoobscura*. There are two convenient measures of the amount of genetic variability present. The first is the fraction of all gene loci which are polymorphic; the second is the fraction of loci in an individual which are heterozygous. Note that the latter fraction is sure to be lower than the former. For example, if 50 per cent of loci were polymorphic with two equally frequent alleles at each locus, only 25 per cent of loci in an individual would be heterozygous; if there were three equally frequent alleles, the fraction of heterozygotes would go up to 33 per cent.

The table on page 185, which is a simplified version of one given by Lewontin, gives values of these measures for those species which have been most extensively studied. The first and most important conclusion is that all the species studied are extremely variable genetically. When it is borne in mind that the electrophoretic technique detects less than one third of all gene differences, the implication of the table is that most loci are polymorphic in most populations. The idea of a 'normal' or 'wild-type' genotype from which there are rare departures is no longer tenable.

On pages 102–6 I described the debate which is going on between supporters of a 'selectionist' and a 'neutralist' interpretation of protein evolution. The same two schools are disputing about the causes of this extensive protein polymorphism. The neutralists argue that most (though not necessarily all) of the observed electrophoretic variability is without effect on the fitness of individuals; the selectionists argue that it is maintained selectively, by heterosis, frequency-dependent selection, or selection favouring different genotypes in different places. I do not think there is as yet any overwhelming reason to take one view or the other, but it is certainly worth reviewing some of the arguments which have been used.

Two main arguments have been used on the neutralist side. The first is that the pattern of variation is just what one would expect if some changes in enzymes have no effect on fitness. This argument requires that the frequencies of alleles at different loci be compared with some theoretically predicted distribution. I do not think this approach has been helpful, if only because the distribution to be expected depends not only on the population numbers now but also on the numbers for many generations in the past, and these we do not know.

The second neutralist argument depends on the concept of 'genetic load'. To explain this concept, imagine a heterotic locus a *Drosophila* with two alleles, such that the fitness of the two homozygotes is only 80 per cent, say, of the fitness of the heterozygote. The two alleles will be equally frequent, so with random mating half the zygotes formed will be heterozygotes and half homozygotes. If selection acts through differences in survival, then at least 10 per cent of the zygotes must die; thus even if all heterozygotes survive, 20 per cent of the homozygotes must die. This is expressed by saying that there is a 10 per cent genetic load associated with this locus. Suppose now that there are 5,000 gene loci in *Drosophila*, and that of these 20 per cent, or 1,000 loci, have heterotic alleles of this kind. Suppose also that selection acts independently on the different loci. Then for each locus a fraction of 0.9 of the population survives, and for all loci the fraction surviving would be 0.9^{1000}, or only one individual out of 10^{46}. Obviously this is absurd; since a female fruitfly is unlikely to lay more than 200 eggs in a lifetime (half of them male) at least one out of every hundred eggs must survive.

The genetic load argument for neutrality amounts to saying that there could not be enough selective deaths to maintain the observed polymorphisms selectively. The argument has been hotly disputed. First, the fitness differences need not be as big as I supposed. If, for example, homozygotes had 99 per cent of the fitness of heterozygotes, the overall fraction surviving would be 0.995^{1000}, or only a little less than one per hundred; this is less absurd though still implausible. A second and perhaps more decisive argument is that selection need not and probably does not act independently on the different loci. To explain this point, the assumption of independence is equivalent to assuming that selection acts successively on the different loci, so that 90 per cent survive selection on the first locus; of these survivors 90 per cent survive selection at the second locus, and so on through the 1,000 loci. An alternative and perhaps more plausible idea is that those individuals heterozygous at the largest number of loci survive, regardless of which loci are involved. Rather surprisingly, with this type of selection a given number of heterotic loci can be maintained with a very much smaller genetic load.

For this and other reasons, the genetic load argument does not seem to me decisive, although it is not as totally irrelevant as some selectionists appear to think. But one thing is clear. If there are large numbers of heterotic loci, the selective advantages associated with each one must be small, and therefore difficult to measure directly. This is an unfortunate conclusion, because it means that the argument is going to be difficult to settle.

I turn now to the arguments in favour of the selective view. The first type of evidence concerns the frequencies of alleles in different parts of the geographical range of a species. If, for example, it could be shown that a particular allele was common in hot places and rare in cold places this would suggest a selective basis for the difference, although as we shall see it would not prove it.

The picture for *Drosophila pseudoobscura* can be summarized as follows. Most of the polymorphic alleles show remarkably similar frequencies throughout the range of the species, despite considerable climatic differences. There are two exceptions to this rule. First, there is an isolated population in the Andes, 1,500 miles from the main distribution; this population is much less variable electrophoretically. Second, there are three loci at which allele frequencies do vary geographically. These are all situated on the third chromosome, which also has extensive inversion polymorphism (see page 186). Prakash and Lewontin have shown that different alleles tend to be associated with different chromosome orders, and that the different allele frequencies can be fully accounted for by different inversion frequencies in different places.

These facts seem to me to fit rather well with the neutralist theory. If the polymorphisms were selectively maintained one would expect allele frequencies to vary with changes in environmental conditions, just as the inversion frequencies do in fact vary. The rather uniform Andean population could be explained if it is descended from a few chance immigrants which did not have the full range of alleles. The third chromosome alleles could themselves be neutral, and vary in frequency because of their tight linkage with other loci within the inversions.

Extensive studies have been made by Ayala on *Drosophila willistoni* and three related species, all widely distributed in tropical America. Ayala finds that in *D. willistoni*, as in *D. pseudoobscura*, allele frequencies are very uniform geographically. The allele frequencies on the Caribbean islands are very similar to those on the mainland, even though island populations are much less polymorphic for inversions. When a comparison is made with the three related species, Ayala finds that the same loci tend to be polymorphic in all species. In some cases the alleles present and their frequencies are very similar in two species; in other cases quite different alleles may be present. I find these observations difficult to explain either on the selective or the neutralist hypothesis.

One last example of the geographical distribution of allele frequencies is worth discussing; it is the study by Allard and his colleagues of the wild oat, *Avena barbata*, in California. This species was accidentally introduced by the Spaniards from Europe. It differs from the species listed in the table on page 185 in reproducing mainly by self-fertilization; only about 1 per cent of seeds are produced by cross-fertilization. Allard found that in the drier parts of California plants are monomorphic at five enzyme loci, whereas in moister regions populations are polymorphic (although because of selfing most individuals are homozygous), the commonest allele at each locus being different from that which is universal in dry areas. These and other observations make it clear that there are two genetically different forms of the wild oat in California, one adapted to dry and the other to less dry habitats. However it is not clear whether the enzyme loci themselves contribute to these adaptations and hence are under selection, or whether they are themselves neutral and are simply acting as 'markers' whereby the two forms can be recognized. The latter possibility arises because the plant is self-fertilizing, so that sets of genes will be held together as if they were tightly linked.

Both Ayala's data on the *D. willistoni* group and Allard's data on *Avena barbata* are easier to interpret on the selective than the neutral hypothesis. Both authors in fact strongly support a selective interpretation. Nevertheless I think it would be wiser to reserve judgement. One other approach is to ask whether there are differences between species in the extent of genetic variability. In general, the table suggests a remarkable degree of similarity between species in this respect. It is worth noting that the Horseshoe Crab, a 'living fossil' which closely resembles in structure animals living 400 million years ago, is not less variable than mice, men or fruitflies. There are, however, differences in variability. For example, Avise and Selander showed that genetic polymorphism was reduced or absent in cave populations of the fish *Astyanax mexicanus*. This is to be expected if cave populations are small, or are established by a few founders, but it does not tell us whether polymorphism outside the cave is neutral or selective. Selander has also pointed out that there is an indication in the data so far available that vertebrates may be less polymorphic than invertebrates; the significance of this difference is far from clear.

One might ask why it is not possible to demonstrate selection acting on enzyme loci, supposing that it does so, by the same population cage techniques used by Dobzhansky to demonstrate selection acting on inversions. Here an experiment on the esterase-5 locus in *D. pseudoobscura* by Yamazaki is illuminating. He worked with two alleles which are common in natural populations. The details of his experiment are important for reasons which will emerge in a moment. Starting from a 15-year old population cage which had 55 per cent of the *S* (slow) and 45 per cent of the *F* (fast) allele, he established 22 lines from single paired matings which were homozygous for the *S* allele, and another 22 lines homozygous for the *F* allele. He used these 44 lines to establish 12 population cages, starting with different frequencies of the *S* allele (from 10 per cent to 90 per cent), on different foods, and at different temperatures. In every case the frequency stayed at its initial value for from one to two years, indicating that no strong selective force was acting on the locus.

Now other experiments show that if one starts, not with 22 lines of each kind, but with only one line of *F* and one of *S* flies, the results are quite different. During the first few generations there is usually a rapid change of gene frequency, showing that selection is acting. The reason for the discrepancy is as follows.

If only a single line of each kind is used, the F and S alleles will initially be linked to different alleles at other loci. Hence selection may be acting not on the enzyme locus itself but at loci linked to it. That this is the case is shown by the fact that when Yamazaki took care to randomize the genes linked to the F and S alleles, no selection could be measured. In interpreting observations on enzyme polymorphism, it is imperative to bear in mind the effects of linkage.

Yamazaki interprets his results as showing either that the esterase-5 alleles are neutral, or that selective differences are too small to be easily measured. One should perhaps mention the third possibility: that they are maintained in the wild by strong selective forces which do not operate in the depauperate environment of a population cage.

There is an obvious drawback to the various methods of investigating selection on enzyme loci so far discussed. It is that there is no evidence from the functioning of the enzymes themselves as to why different alleles should be favoured in different circumstances. Thus if one were to suggest that the reason why the frequency of the gene for melanism in the Peppered Moth has increased in industrial areas is that it is closely linked to some other unidentified locus on which selection is acting, the suggestion would rightly be laughed out of court. This is because the causal connection between colour and survival is understood. Clarke has recently argued that a demonstration of selection acting on enzyme variants will have to start from a study of the properties of the different forms of the enzyme; in taking this position he is very much in the tradition of the English school of 'ecological genetics' which stems from the work of E. B. Ford, and which has been responsible for the work on industrial melanism and on colour patterns in *Cepaea* described in the last chapter.

The enzyme polymorphism which has been investigated most actively from this point of view is that for alcohol dehydrogenase (ADH) in *Drosophila melanogaster*. Almost all natural populations are polymorphic for a fast (F) and a slow (S) allele. Gibson showed that the enzyme produced by the F allele has a higher activity than that produced by the S allele, but is less stable at high temperatures. As these observations would lead one to predict, the F allele increases in frequency in laboratory populations kept on a medium with a high alcohol content. It therefore seems very unlikely that this polymorphism is selectively neutral in the wild, although it is not yet clear how selection maintains a balanced polymorphism, instead of fixing different alleles in different habitats. There is no reason to think that these results on ADH are atypical; for example, Harris has reported that of 23 enzyme polymorphisms in man which have been examined, biochemical differences have been demonstrated in 16 cases.

CHAPTER 12

Altruism, Social Behaviour, and Sex

Altruism, defined by the Oxford English Dictionary as 'regard for others, as a principle of action', might seem at first sight to have little to do with evolution theory. Yet there are occasions when animals (and even plants – see the 'corky' syndrome, page 273) do things which increase the chances of survival of other members of the species, at the expense of reducing their own chances. I shall call such actions altruistic, without wishing to imply that the performer has a conscience or an ethical philosophy. The difficulty for the theory of natural selection is obvious: if it is survivors who transmit their characteristics to future generations, how can altruism be established?

One answer is that in evolution it is not the survival of the individual that matters but of the offspring of that individual. Thus a lapwing which feigns injury when its nest is threatened by a hawk is acting altruistically. But the genes responsible for injury-feigning will be maintained by natural selection, because the chances that a parent will be killed by a hawk are increased only slightly by feigning injury, whereas the chances that its offspring, who inherit their parent's genes, will survive are increased appreciably. Injury-feigning is a more striking example of parental care than most, but it does not require any special explanation.

Yet there are patterns of behaviour which increase the survival chances of individuals other than the offspring of the performer. This may often be because the performer does not distinguish between situations in which the beneficiaries are its own offspring, and those in which the beneficiaries are not. For example, the protective responses shown by female mammals to the young of their own and often of other species probably evolved because the most efficient way of ensuring that a female will protect her own young is that she should evolve an instinctive response to certain very generalized features of young mammals. Such undiscriminating responses can lay a species open to exploitation by others; for example, many birds are exploited by cuckoos, ants are exploited by numerous species of insects, and women are exploited by lapdogs.

An absence of discrimination is important in the evolution of most altruistic characteristics, but does not by itself explain all of them. For example, many small passerine birds give a special alarm note when they see a flying predator. Birds hearing such an alarm note stay still, and so are more likely to escape detection. Such notes may first have been given by parents to their young during the breeding season, but they are today given by members of winter flocks, which do not consist of parents and their children. Birds could surely have evolved the capacity to give the alarm note in the breeding season but not in the winter, so it follows that the alarm note is given in winter flocks because the habit is favoured by selection.

Alarm notes are high-pitched and on a narrow range of frequencies; such notes are difficult to locate. It could therefore be argued that a bird giving the alarm does not risk its life. But in fact the peculiar features of the alarm note point the other way. Adaptation is explained by natural selection; in this case, alarm notes are today difficult to locate because birds which in the past gave more easily locatable notes were killed by predators. In other words, it is dangerous to give an alarm note.

Thus there are reasons to think both that selection does maintain the habit of giving an alarm note in winter flocks, and that there is a risk attached to giving it. How are these apparently contradictory conclusions to be reconciled? If members of winter flocks are unrelated, I do not think they can be. But if flocks tend to contain brothers and sisters, or cousins or other close relatives, then an explanation can be given. Let us suppose that an individual gives the alarm because he carries the gene A; if he carried gene a he would selfishly remain silent. By giving the alarm he helps to save the lives of other members of the flock. If the other members are related to him, then they will tend to have the gene A. So his action in giving the alarm tends to preserve the gene A in his relatives, but to eliminate it in himself. Whether the gene A or a will increase in frequency depends on the balance between the risk to the individual and the advantage to the flock. The mathematical treatment is difficult, but it is easy to see that the more closely are the members of the flock related, the greater the risk an individual will run to confer a given advantage to the other members of a flock. If the flock is merely a random sample of the whole population, selection will not favour the running of any risk at all.

I have suggested the term 'kin selection' for the process whereby a characteristic is established because of its effects on the survival of the relatives of its possessor. The process of kin selection was clearly understood by Haldane, as is shown by his remark that he was prepared to lay down his life for two brothers or eight cousins; recent interest in the process, and appreciation of its importance in the evolution of social behaviour, is due mainly to Hamilton. Two other examples will be given. One concerns the differences in the length of post-reproductive life of different species of moth, according to whether the species is cryptically coloured and palatable, or brightly coloured and distasteful. Blest has pointed out that, if kin selection is effective, one would expect cryptic adults to die immediately after reproduction, since the fewer cryptically coloured individuals there are about, the less chance a predator has to learn to recognize them. But in a brightly coloured and distasteful species, an individual would confer an advantage on its relatives by living a long time after reproduction, thereby increasing the chance that a predator will attack it and learn that it is distasteful. Blest has confirmed this idea by some observations on tropical saturniid moths. He found that the cryptic species lived for a shorter time than those with warning coloration, partly because they become more excitable with age, and end by flying continuously until they die. From an evolutionary point of view this continuous flight is a form of altruistic suicide.

Another example of kin selection is the evolution of sterile castes in the social insects. A worker bee is a sterile female who spends her life looking after her sisters, some of whom will be fertile queens. The difference between a worker and a queen is caused by the kind of nutrition given to the grubs. But the capacity to be sterilized by a particular diet is itself genetically determined, and its evolution must be explained by kin selection. Hamilton has pointed out an entertaining twist to the story. Social life has been evolved on at least four separate occasions by the insects—by the ants, bees, wasps, and termites. The first three of these four groups belong to the same insect order, the hymenoptera. What features of the hymenoptera have predisposed them to evolve societies? Hamilton suggests that the feature in question is their 'haplo-diploid' genetic mechanism. Among the hymenoptera, solitary or social, males develop from unfertilized eggs, and so are haploid (i.e. have only a single set of chromosomes); females develop from fertilized eggs and are diploid. As shown diagrammatically in Figure 22, a female of a haplo-diploid species has three-quarters of her genes in common with her sisters but only half her genes in common with her daughters, whereas in a normal diploid species she has half her genes in common both with her sisters and her daughters. Thus a female hymenopteran does more to preserve her own genes if she stays at home and looks after her sisters than if she goes out and starts a family of her own.

If kin selection were relevant only to a few peculiar problems such as the evolution of alarm notes or of social hymenoptera, it would perhaps be no more than an amusing gloss on evolution theory. But there are two phenomena of great importance, namely the evolution of mechanisms regulating population density and the evolution of sex, which may require kin selection or something like it to explain them.

If the numbers of a sexually reproducing species are to remain constant, each pair must on the average produce two offspring which survive to sexual maturity. Even a slight departure from this average, if continued for many generations, would lead to an enormous increase or decrease in numbers. Yet most species maintain a more or less constant density year after year, or maintain a density which fluctuates fairly regularly between certain limits. This can be explained only if the fertility or chances of survival of individuals decline as the density

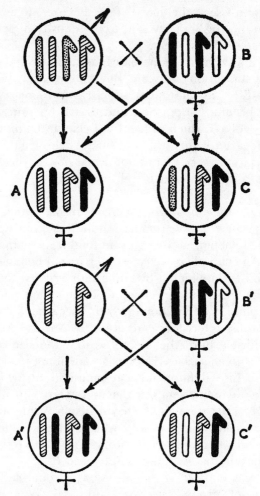

Figure 22. Why are the hymenoptera social? Above, in a normal diploid species, female A has exactly half her genes in common with her mother, B. In common with her sister C she has on the average half the genes she inherited from her father, and half of those from her mother.

Below, in the haplo-diploid species, female A' has exactly half her genes in common with her mother B'. In common with her sister C' she has all her father's genes, and, on average, half her mother's; in all, three-quarters of her genes.

increases, and rise as the density falls. Factors affecting fertility or mortality in this way are said to be 'density dependent' the problem is to identify them.

Clearly if a population increases too far, its numbers will be reduced by starvation or disease. But it has recently been argued by Wynne-Edwards that populations rarely increase so far that starvation becomes important, and that behavioural mechanisms have evolved which prevent a species outrunning its food supply. He argues further that since such behaviour is advantageous to the species, but not necessarily to the individual, it requires a special kind of selection, which he calls group selection, to account for it.

There are some behavioural mechanisms which may well help to prevent excessive population increase, but which could evolve by selection acting on individuals, without group selection. An example is the type of territorial behaviour shown by many birds, in which a breeding pair occupy and defend a territory in which they collect food for their young. Such behaviour limits population growth, because a bird which cannot establish a territory cannot breed. Natural selection will tend to adjust a bird's behaviour so that the size of territory typical for a given species contains an adequate food supply for the young. This will happen because individuals which are too aggressive, or which attempt to defend too large an area, will be likely to get hurt, or will waste on display time and energy needed for raising their young, whereas birds which are too timid will fail to establish a territory, or will establish one too small to contain an adequate supply of food.

In this case, there seems no need to invoke a special mechanism of group selection, although Wynne-Edwards would not agree with this conclusion. He also believes that a number of complex group displays have evolved because they provide necessary information to members of the species about population density. Such displays are necessary because, if density is to influence breeding behaviour, there must be some way in which individuals can become aware of it. Indeed Wynne-Edwards goes so far as to suggest that the origin of all social behaviour lies in such 'epideictic' displays, whose function it is to provide information about population density.

These views have aroused considerable interest and controversy, which has centred mainly on the problem of group selection. The difficulty is best explained by considering a particular case. In many small rodents, when the population density rises there is much fighting and some of the adult mice stop breeding; in lemmings, many individuals undertake what seems to be a suicidal migration. Now this may merely be a case of discretion being the better part of valour; it may be better to run away or to be submissive than to get killed or hurt. Even a migrating lemming may occasionally find a new promised land. If so, individual selection can account for the behaviour, and no difficulty arises. But it is Wynne-Edward's argument that the evolutionary reason why individuals refrain from breeding is that by doing so they ensure that the population will not outrun its food supply. Submissive individuals favour the survival of the group to which they belong. The difficulty is of course that the genetically 'altruistic' individuals – the ones which refrain from breeding for the common good – do not pass on to the next generation the genes by virtue of which they are altruistic.

There may be ways out of this difficulty. Clearly a group consisting entirely of altruistic individuals will do better than one consisting entirely of selfish ones, since the latter will first increase without limit and then starve. The problem is to explain how a group comes to consist wholly of altruistic individuals in the first place, since in a mixed group altruism will be eliminated by selection. There are two possible answers. The first, which applies to our own species and perhaps to the more intelligent social mammals, is that the difference between altruism and selfishness may be a matter of education and not of genetics, so that altruism may be spread by education to all members of a group. This requires that all members of the species be genetically educable, and that some method exists of eliminating genetically ineducable or amoral individuals should they arise by mutation. Such elimination is likely to result partly from a genetically determined and instinctive intolerance of non-conformism, and partly because amoral individuals are unlikely to care properly for their own children; it may be difficult for selection to produce an individual which will sacrifice itself for its relations but not for strangers.

A second method of establishing wholly altruistic groups arises if the groups are numerically very small, or are occasionally reduced to one or a few individuals. In such cases, a wholly altruistic group can arise by chance. If a population is divided into a number of groups, between which little interbreeding takes place and which are periodically reduced to small numbers, then a genetically determined tendency to refrain from breeding at high population densities could spread by group selection. But the circumstances in which group selection can operate are unusual. In most cases in which behaviour is important in limiting population density, that behaviour has probably evolved by individual selection.

The other problem to which group selection may be relevant is that of the origin and maintenance of sexual reproduction. We are so accustomed to associating the ideas of sex and of having children that it is easy to forget that, at the cellular level, the sexual process of fertilization is the precise opposite of the reproductive process of cell division; fertilization is a process whereby one cell is produced from two. It is, in fact, easy to see that in the short run sexually reproducing females will be at a twofold selective disadvantage compared to parthenogenetic females. Thus, suppose a population consists of a mixture of sexual and parthenogenetic females, the former producing equal numbers of male and (sexual) female offspring, and the latter only parthenogenetic females like themselves. If the two kinds of female lay equal numbers of eggs, and if survival probabilities are equal, then the parthenogenetic type will have a twofold selective advantage, and will increase in frequency very rapidly. Sexual reproduction means that a female wastes half her energy producing males.

There are some situations in which this advantage of asexual reproduction does not operate:

(i) If both parents care for the young, the sexually produced offspring will have better chances of survival. Males are no longer a waste.

(ii) In most single-celled organisms and in some simple multicellular plants, there is no differentiation of the gametes into large immobile eggs and small mobile sperm.

In such organisms, sexual reproduction is not an immediate disadvantage. Since sexual reproduction presumably arose among eukaryotes before there was any differentiation of the gametes, we do not have to take this twofold disadvantage into account when discussing the origin of sex, but only when discussing its maintenance during subsequent evolution.

(iii) In many higher plants, seeds and pollen are produced by the same individual, and some animals (e.g. most snails and flatworms) are hermaphrodite, the same individual producing both eggs and sperm. In these cases no material is wasted on males. There is, however, still a difficulty. Why should an individual accept foreign sperm or pollen to fertilize its egg cells, since it could increase the number of its own offspring by self-fertilization?

In most situations, then, there are short-term disadvantages associated with sexual reproduction. The compensating advantage of the sexual process is that it increases the range of potential variation in a population, and therefore its evolutionary plasticity. The vegetative progeny of an individual (i.e. progeny produced without meiosis and subsequent fertilization) are genetically identical with each other and with their parent, unless a mutation has occurred. Suppose that among the vegetative descendants of a single individual ten different mutations have occurred, at different loci and in different lines of descent. Then there will exist ten different genotypes upon which selection can act. If however, these ten lines could cross sexually, three different genotypes would be possible at each locus (AA, Aa, aa), and these could be combined in any manner, giving a total of $3^{10} = 59,049$ different possible genotypes. To put the matter another way, if different favourable mutations occur in different individuals of a vegetatively reproducing species, there is no way in which they can be combined in a single individual.

Thus the sexual process is a means of ensuring evolutionary plasticity at the expense of interfering with reproduction. Many species, both of animals and plants, have evolved ways of getting the best of both worlds. In single-celled organisms, fertilization necessarily interrupts cell division, but it is often found to occur only when food shortage or other conditions would in any case slow down cell division. Many plant species, in addition to forming seeds by a sexual process, can also reproduce vegetatively, by stolons, tubers, bulbs, etc. Vegetative reproduction by budding is also not uncommon among invertebrate animals. Some insect species, such as aphids (e.g. greenfly), have a series of generations during the summer consisting solely of parthenogenetic females, but in autumn both males and females are produced parthenogenetically, and mate, the result of this sexual cross being the production of a new generation of parthenogenetic females.

Now if the advantage of sexual reproduction is that it increases the range of potential variation in a population, then the advantage refers to the population as a whole, and not to any particular individual in it. It follows that sexual reproduction has been established as the rule, both in animals and plants, because selection has favoured some populations at the expense of others. Those populations which could evolve most rapidly to meet changes in the environment have survived. We are again faced with the problem of group selection, in which the unit selected is the population and not the individual.

Some biologists have argued that sexual reproduction can be explained without invoking group selection. G. C. Williams, the strongest proponent of this view, has pointed out that in groups such as the aphids, in which the sexual phase of the annual cycle could presumably be eliminated rather easily if it were selectively advantageous to do so, there must be some short-term advantage in sex. He suggests that this advantage lies in the production by an individual sexual female of a more variable progeny. Suppose that in a given patch of environment a sexual female contributes, say, 50 'propagules', all genetically different, and an asexual female produces 100 propagules, all identical to their parent. If selection is very intense, only that propagule which is genetically best adapted to the particular patch will survive; imagine if you like that the propagules are seeds of a forest tree, falling in an opening where an old tree has fallen and left space for one new mature tree. This best-adapted

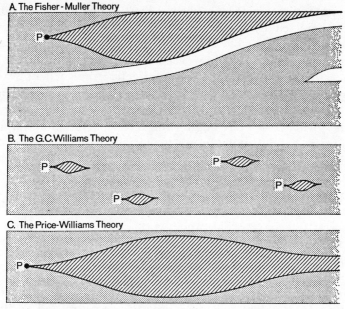

Figure 23. Three theories of the selective maintenance of sex. Each band represents a population or species, with time travelling from left to right. P represents the occurrence of a parthenogenetic mutant, and the hatched area its descendants. A, the proposal of R. A. Fisher and H. J. Muller, according to which those species which abandon sexual reproduction go extinct; B, the proposal of G. C. Williams, according to which sex is an advantage within a population in the short run; C, a suggestion by Mary Williams and G. R. Price.

propagule is likely to be one of the sexually produced ones, since all the asexual ones are identical. The asexual parent is like a person who buys 100 tickets in a raffle with only one prize, and finds that all his tickets have the same number on them; the sexual parent buys fewer raffle tickets, but they all have different numbers. In this process, sex confers an immediate advantage. The idea is an attractive one, although there are real difficulties about accepting it.

These two theories about the selective maintenance of sex are illustrated in Figure 23. In the same figure a third, intermediate theory is shown. So far as I know, this third theory has not been published; it has been suggested to me independently by Dr

Mary Williams and by Dr George Price. The idea is that when
a parthenogenetic mutant first arises in a population, it does
indeed have a twofold selective advantage, and increases in
frequency. It would, however, take perhaps hundreds of
generations before the mutant completely replaced its sexual
competitors. During this period, the sexual population would
continue to evolve, whereas the asexual one would evolve far
more slowly. If so, there might come a time, before the complete
elimination of the sexual form, when its increase in fitness would
more than counterbalance the twofold disadvantage of sex;
from this point on, the asexual form would decrease in
frequency. Dr Larry Gilbert has pointed out to me that this
argument is reinforced if one remembers that a species usually
has a wide geographical range, with locally adapted races. A
parthenogenetic race might have a twofold advantage at its
point of origin, but be unable to compete successfully over the
whole geographical range of the sexual species.

It is fairly certain that fertilization and meiosis originated
among single-celled organisms, and were transmitted by them
to their many-celled descendants. These processes are complex
ones, and it is difficult to imagine that they can have evolved
rapidly, let alone as a result of a single mutation. It is equally
difficult to see how the habits of nuclear fusion and of meiosis
could have spread through a population if the only advantage
they conferred was the long-term one of increasing the future
evolutionary resources of that population. We are therefore
driven to the conclusion that the early stages in the evolution of
the sexual process took place under the influence of selective
forces quite different from those which were responsible for the
maintenance and spread of sexual processes once they were
perfected.

It is interesting to speculate on what these selective forces may
have been. It was suggested earlier (page 118) that the enzymes
required for genetic recombination between chromosomes may
have evolved in the first instance to repair damaged DNA, and
that genetic exchange may have arisen very early in the history
of life. Once the cellular level of organization was achieved,
genetic exchange required that DNA from different cells be

brought together in a single one. Among the prokaryotes this is not brought about by cell fusion; instead, DNA fragments (or in some cases whole chromosomes) from one 'parent' are introduced into the other, in ways which seem to have more in common with the process of viral infection than they do with sexual processes in higher organisms.

In contrast, the eukaryotes have evolved the processes of meiosis, whereby a diploid cell gives rise to haploid gametes, and fertilization, whereby two gametes fuse to form a new diploid cell. These more elaborate and regular sexual processes seem necessary if genetic recombination is to take place between eukaryotic cells, with their multiple chromosome sets and regular mitotic divisions. How did meiosis and fertilization evolve? They do not appear to be elaborations of sexual processes found among prokaryotes. In earlier editions of this book, I suggested that cell fusion may have originated as cannibalism, one cell swallowing another. I no longer think much of this suggestion, although either cannibalism or parasitism may have been the origin of the symbiotic events which gave rise to the eukaryotes (see pages 118–20). It seems more likely that sexual fusion evolved from the type of fusion which can be observed today among many fungi, in which two individuals fuse to form a single organism with two genetically different kinds of haploid nuclei; such an organism is known as a heterokaryon. Such fusion would confer a selective advantage similar to that conferred by hybrid vigour. Thus different vegetatively reproducing lines would accumulate different types of genetic defect by mutation. The fusion of such lines would lead to a recovery of vigour for much the same reason as the crossing of inbred lines leads to such a recovery.

The next stage would be nuclear fusion, with the two sets of chromosomes being attached to the same spindle during mitosis. This would have the advantage of preserving a balanced hybrid constitution indefinitely, instead of permitting one nuclear type to multiply more rapidly than the other. The real difficulty lies in understanding why such an asexual diploid organism should evolve the process of meiosis. In the long run meiosis and fertilization may be favoured because they make possible more rapid evolution, but this cannot account for the origin of meiosis, which remains one of the unsolved problems of biology.

The division of a population into males and females is not a necessary feature of sexual reproduction. For example, in *Paramecium aurelia* individuals belong to one of two 'mating types', which can be distinguished only by the fact that members of one type will 'conjugate' only with members of the other. Yet there is a process of meiosis, after which an individual contains two identical haploid nuclei equivalent to the nuclei of gametes. In conjugation, two individuals of different mating types lie side by side, and one of the two nuclei from each of them is transferred to the other, so that each then contains two haploid nuclei, one derived from each 'parent'. These two nuclei then fuse to form a new zygotic nucleus. The genetic results of this process are very similar to the results of the more familiar methods of sexual reproduction in higher organisms, yet there is no differentiation between the sexes.

A characteristic feature of many-celled animals is the division of labour between the different cells of the body. There is a corresponding division of labour between the two different kinds of gamete. One kind, the ovum' or egg cell, is relatively large and carries the necessary food reserves; the other, the sperm, is small, and possesses a special organ, the flagellum, to enable it to reach the egg. In hermaphrodite animals, sexual differentiation extends no further, all individuals being alike in producing both kinds of gametes. In animals in which the sexes are separate, there may still be little difference between males and females, other than that the former produce sperm, the latter eggs. This is often the case in marine animals in which fertilization is external, as, for example, in echinoderms (starfish and sea urchins) and in most bony fish.

With the evolution of internal fertilization, differences have arisen between adult males and females which parallel those between the gametes which they produce; females have evolved structures, such as the placenta and mammary glands of mammals, concerned with the nutrition of the growing embryo, whereas males have evolved structures which can introduce sperm into the female, and sometimes possess specially developed sense organs or powers of locomotion to seek out the females. However, such a division of labour, in which it is the female which is primarily responsible for providing the young with food and protection, is far from universal. In many bird species the labours of nest-building, incubation, and feeding the nestlings are shared equally by the two sexes. In some animals the protection of the young is carried out by the males. It is the male stickleback which builds and protects the nest, the male emperor penguin which incubates the eggs, and in the European toad *Alytes obstetricans* the male winds the strings of eggs around his body, where they remain until the tadpoles are ready to hatch.

Before discussing the selective forces responsible for the evolution of sexual differentiation, something must be said of the mechanisms determining the sex of individuals In a few animals. the sex of an individual is not determined at birth, but by the environmental conditions in which it is reared. In most species, however, sex is determined genetically. A variety of such mechanisms are known, but perhaps the commonest is the X-Y mechanism, found in mammals and birds, and also in many insects, including *Drosophila* (see page 62).

Once a difference between males and females has evolved, much of the further elaboration of sexual organs and behaviour can result from selection acting on individuals. There are, however, certain structural and behavioural characteristics of the two sexes which depend for their effective functioning on a proper coordination between a male and female; in such cases a change in one individual may be a selective advantage only if appropriate changes also take place in others. An example will make this point clearer. In many marine animals, for example oysters and sea urchins, fertilization is external, the sex cells being shed into the sea water where fertilization takes place. An individual is more likely to leave offspring if it releases its gametes at the same time as do other individuals in the neighbourhood. In a number of cases it has been shown that an individual shedding its gametes also releases into the water a chemical substance which stimulates other individuals to do likewise. The difficulty in explaining the origin of such a situation is this: the individual which first produces such a chemical substance will not be at a selective advantage unless other individuals are stimulated by it, whereas individuals are unlikely to respond to a substance which has not previously been produced. It is no good making a signal unless it is understood, and a signal will not be understood the first time it is made. This is true whether the signal is a chemical substance, a characteristic movement or sound, or the display of some coloured or otherwise ornamented structure.

The probable answer to this difficulty is as follows. There are many constituents of semen other than sperm, and these do not necessarily have any signalling function; they may, for example, be important for the nutrition of the sperm. Probably those substances which now have a signalling function originally served some other purpose; that is to say, individuals producing such substances were at first favoured by selection, not because other individuals were thereby stimulated to release their sex cells, but for some other reason. Once, however, such a substance was the normal accompaniment of the release of gametes, any other individual which responded to the presence of that substance by releasing its own gametes would be at an advantage. The evolutionary process may therefore have been as follows: first, some substance, although not functioning as a signal, may normally be released with the gametes; second, selection will favour any individuals which respond to the presence of this substance by releasing their own sex cells; finally, once the substance has acquired a signalling function, individuals will be selected which produce more of the substance, or perhaps which produce a slightly different substance, more easily perceived and therefore more effective as a signal.

The same argument can be applied to movements or to patterns which act as signals in courtship. Tinbergen has pointed out that animals, particularly when in a state of stress or inner conflict, often make movements which are inappropriate to their immediate circumstances, although well suited to others; examples from human behaviour include scratching one's head when puzzled, straightening one's tie when nervous, and a whole range of other fidgets. Such movements have been called 'displacement activities'; their relevance here is that they may acquire a significance in communication, just as was suggested above in the case of specific chemical substances. Thus movements originally appropriate to nest-building, preening, or feeding the young have come to play an important part in bird courtship. Once such a movement has acquired the function of a signal (i.e. once other individuals have come to recognize it as an indicator of the physiological state of the individual making the movement, and to respond to it in an appropriate way), then there will be a selective advantage in further elaborating the movement, and in the evolution of particular patterns or colours which render it more conspicuous. This process has been called 'ritualization'. It has been important in the evolution of all kinds of behaviour involving communication between different individuals. This may take place not only between members of the same species, as in courtship, territorial behaviour, or the coordination of flocks of birds or schools of fish, but also between members of different species, as in the evolution of the 'warning' coloration of distasteful animals.

In the evolution of mating behaviour, it is possible to recognize three kinds of selection pressure which may operate:

(i) Selection ensuring that an individual shall mate, or increasing the frequency with which it mates.

The release of chemical signals by marine organisms with external fertilization is an example of this kind of selection. An important component of the courtship of higher animals has evolved to elicit a sexual rather than an aggressive response from the partner. For example, the usual reaction of a female spider to an animal slightly smaller than itself is to attack and eat it. Consequently male spiders are in danger of being eaten by their prospective spouses, and have evolved complex patterns of behaviour to avoid this eventuality; for example, the male may vibrate the web made by the female in a characteristic manner. Similarly, one of the movements in the courtship of the Black-headed Gull has been explained by Tinbergen and Moynihan as an 'appeasement' ceremony. These gulls defend a breeding territory around their nest, displaying at intruders by holding the head lowered with the beak horizontal and pointing at the intruder. During one phase of courtship the two partners stand with their heads raised, the beak pointing downwards and the head turned away from the partner, thus offering an aspect contrasting as sharply as possible with that shown in the threat display.

A particularly high degree of elaboration of the sexual characters may evolve in those species in which a male can mate with a number of females (polygyny). This is commoner than polyandry, in which a single female mates with a number of males, because the number of offspring produced by a female is usually limited by the number of eggs she can lay and not by the amount of sperm she receives, whereas a male which can mate twice may thereby double the number of offspring it leaves. In mammals, polygyny is common among ungulates (cattle, antelopes, deer, etc.) but rare among carnivores. In carnivores, monogamy, at least during the breeding season, makes it possible for both partners to help in feeding the young, whereas young ungulates must depend either on their mother's milk or on plant material they have collected for themselves. Consequently a male ungulate, once mated, can do little further to increase the chances of survival of its offspring, and so natural selection has emphasized those characteristics, such as horns and antlers, which increase their chances of mating more than once. It is common to find that male ungulates are larger and better equipped with offensive weapons than are females, whereas in carnivores there is seldom any appreciable difference in size between the sexes.

(ii) Selection ensuring that an individual shall mate with a member of its own, and not of a related, species.

Evidence that selection of this kind occurs, and is important in the process of speciation, will be discussed in later chapters.

(iii) Selection ensuring that an individual will mate with one member of its own species rather than with another. For example, a female which mates with the first male of her own species she comes across may satisfy conditions (i) and (ii) above, yet her fitness may be further increased if she shows greater discrimination, and chooses one of a number of possible males as a mate, the male chosen being more fertile than the average.

This was the kind of selection which Darwin had in mind when writing of sexual selection. He pointed out that in a species with equal numbers of males and females, and one which, at least for a single breeding season, is monogamous, all individuals have the opportunity to mate.[1] Therefore, although a male which possesses particularly striking sexual characters may thereby be enabled to be one of the first to find a mate, this will not increase his fitness unless it also ensures that he will mate with a female who is particularly fit as a parent. Similarly, a female who selects as a mate a male with striking sexual characters will not therefore leave more offspring unless the male is also particularly fit as a parent. In other words, in a monogamous species secondary sexual characters will not be perfected by selection unless those individuals with such characters particularly well developed are themselves fitter than the average as parents, and also are able to find mates which are fit as parents. Darwin thought that such an association between well-developed secondary sexual characters and fitness as a parent would in fact exist, because both would be features of the most vigorous members of a population; he also suggested that males with well-developed sexual characters would, on the average, tend to mate with the fittest females, since they would tend to mate with the first females ready to breed, and such females are likely to be the healthiest.

Darwin's ideas on sexual selection have received little attention from later biologists. In no case has it been demonstrated that such selection occurs in a wild population; this is

---

[1] It is usual to find, even in such species, that some mature individuals do not in fact mate; this is certainly true of our own species. If so, it is possible for selection of kind (i) above to cause the evolution of striking sexual characteristics. But Darwin was probably right in thinking that selection of kind (iii), involving conflict between males for particular females, or the choice by females of particular males, is important to monogamous species.

perhaps not surprising, since it would be necessary to show, not only that females are selecting as mates some kinds of males in preference to others, but also that, by so choosing, females are increasing the average number of offspring they leave. I shall therefore describe some observations on laboratory populations of the fruitfly *Drosophila subobscura*, in which both the above facts have been established, and from which it has also been possible to discover the method whereby the females make their choice of mates.

In *D. subobscura*, a single male can mate many times, but females are effectively monogamous. After mating, females store sperm in a special receptacle, releasing a small quantity to fertilize each egg as it passes down the oviduct. Once a female has been inseminated, she will not mate again, not even if she subsequently exhausts all the stored sperm in her receptacle. Consequently, it is possible for an old female to continue to lay unfertilized eggs because she no longer carries any stored sperm, and yet to refuse to mate again.

When two groups of females of similar genetic constitution were mated, one group to outbred males and one to inbred males, and the eggs which they laid for the rest of their lives collected, it was found that the total number of eggs laid by females of the two groups did not differ, but that the proportions of those eggs which hatched were very different. Females mated by outbred males laid an average of over one thousand hatching eggs per female, whereas females mated by inbred males averaged only 260 fertile eggs per female. This was because inbred males produced fewer sperm than did outbred males, and because some of the sperm they did produce were defective.

Since females normally mate only once, it follows that females which mate with outbred males leave about four times as many offspring as do those which mate with inbred males. There would therefore be strong natural selection in favour of females which tend to mate with outbred rather than inbred males when given a choice. Now it was found that if a virgin female was put in the same container as an outbred male, mating took place within one hour in over 90 per cent of cases, whereas if similar virgin females were paired with inbred males, mating took place in only about 50 per cent of cases. It does not follow from these results that there was any process of 'selection' on the part of the females, since they could equally well be explained if a large proportion of the inbred males did not court or attempt to mate.

Direct observation of the pairs showed that the latter explanation is untrue. In many cases an inbred male courted continuously, and even made repeated attempts to mount an unwilling female. Therefore some other explanation of the different proportions of matings in the two cases must be found; this requires a brief description of the courtship behaviour in this species.

The typical sequence of events after placing a virgin female with an outbred male is as follows. After a few minutes in the same container, the male appears to catch sight of the female; he turns and approaches her, giving a series of rapid flicks with his wings. After tapping her with his front legs, he circles round so as to stand facing her head to head, with his proboscis (i.e. licking mouth parts) extended towards her. The female then executes a rapid side-stepping dance, moving first to one side and then to the other, the male side-stepping as well so as to keep facing her. The female then stands still and the male circles round rapidly and mounts. The whole process, from the first approach of the male to the actual mounting, may take only a few seconds.

Sometimes, however, the female may break off in the middle of the side-stepping dance, and turn her back on the male, or fly away, in which case the male will again approach her head to head if opportunity arises. Now if the male is an outbred one, mating usually takes place after one or after relatively few dances. With an inbred male, on the other hand, a whole series of dances may take place, after each of which the female moves away without mating. After such a series of rebuffs, a male may approach a female from the side or from behind and attempt to mount without the preliminary dance, but such attempts are never successful.

These facts suggest that there must be some difference between the behaviour of inbred and outbred males during the courtship dance, which determines whether a female will stand still and accept the male, or will move away. The difference is probably this: the movements of a dancing female are very rapid, but an outbred male will usually manage to maintain his position facing the female. Inbred males on the other hand often lag behind, failing to keep up with the female. It seems very likely that a female will accept a male which has been facing her while she executed her side-stepping dance, but not a male which has been lagging. The difference between the two kinds of males probably arises because of the greater athletic ability of outbred males. This difference is only detected by a female because she dances; dancing is a fly's way of being 'hard to get'.

If this interpretation is correct, a female is not selecting a male because he is fertile, but because he can co-ordinate his movements with hers during the dance. In these laboratory experiments only two kinds of male were used, outbred and highly inbred, and the former were superior not only in athletic ability but also in fertility. If in a wild population only these two kinds of males existed, those females which executed a dance would be at a great selective advantage over females which did not, because they would tend to mate with fertile males and so leave more progeny. Consequently the evolution of the courtship dance could be explained by Darwinian sexual selection.

Selection can only work in this way, however, if those males with characteristics, in this case dancing ability, which increase their mating success are also fitter as parents. Such correlations may well exist because, in Darwin's words, 'both will be features of the most vigorous members of the population'. The correlation was exaggerated in the experiments described by using only outbred and highly inbred males. It would be very difficult to demonstrate the existence of such a correlation in a wild population, but it probably exists. This example has been described in some detail because it probably illustrates a common phenomenon. The movements of animals during courtship, and the structures displayed by such movements, are often elaborated to a remarkable degree. In many species, particularly those in which one or both sexes are monogamous, it is difficult to account for such elaboration in terms of the first two kinds of selection pressure mentioned above, whereas selection of the third kind, ensuring that an individual shall mate with one member of its own species rather than with another, can explain any degree of elaboration of secondary sexual characters.

CHAPTER 13

What are Species?

In earlier chapters of this book an important question has been glossed over. Although it has been emphasized that no two members of a species are exactly alike, and that there are statistical differences between populations of the same species from different places, it has been assumed that animals and plants can be classified into a series of distinct species. To put it another way, it has, for example, been assumed that we can say of any snail collected from any part of the world that it either does or does not belong to the species *Cepaea nemoralis*, or that we can say that one population of flies belongs to the species *Drosophila melanogaster*, and another to the species *Drosophila pseudoobscura*. Clearly this method of naming is possible only in so far as there are discontinuities in the variation of animals and plants, such discontinuities marking the boundaries between one species and another. In this chapter I shall discuss how far it is true that such discontinuities exist.

However, the discussion will be easier to follow if two general points are made at the outset The first is as follows. The theory of evolution holds that existing animals and plants have originated by descent with modification from one or a few simple ancestral forms. If this is true, it follows that all the characteristics by which we can classify them into species have been and are changing, and further that on many occasions in the past a single population has given rise to two or more populations whose descendants today are sufficiently different from one another to be classified as different species. Now there is no reason to suppose that either the processes of modification in time, or the processes of division of a single species into two, have always, or even usually, occurred in a series of sharp discontinuous steps. Therefore any attempt to group all living things, past and present, into sharply defined groups, between which no intermediates exist, is foredoomed to failure. Historically, of course, this argument was reversed; the observation that animals and plants cannot satisfactorily be divided into distinct species contributed to the spread of evolutionary views among biologists.

The second general point is this: even though it is not possible to devise a fully consistent classification into species, the attempt must for practical reasons be made. For example, a few years ago I was studying the ways in which birds are adapted to different kinds of flight – soaring, gliding, flapping, and so on. For this purpose it was desirable to know the weight, wing span, and wing area of as many different kinds of birds as possible. Unhappily bird taxonomists usually do not measure any of these things; for example, they prefer to measure the length of the wing from the wrist to the tips of the primary feathers, rather than from wing-tip to wing-tip, because it can be done more accurately. In fact, the only place where I could find a large collection of the kind of measurements I wanted was in a charmingly entitled paper, 'The first report of the bird construction committee of the Aeronautical Society of Great Britain', published in 1910. But the authors of this paper were unaware of the desirability of giving the scientific names of the specimens measured. Many birds whose measurements were given were identified by the single word 'hawk', which might have meant the hovering Kestrel, the soaring Buzzard or the fast-flapping Sparrow-hawk. Had the scientific name of each species been given, the list would have been of far greater value. The point of this anecdote is obvious. Despite the unavoidable imperfections of any system of classification, an internationally accepted system of naming does enable biologists to convey a fairly precise idea of the kind of animal or plant they have observed by the use of specific names, such as *Falco tinnunculus* for the Kestrel or *Accipiter nisus* for the Sparrow-hawk.[1] In many

---

[1] Since writing this, I have learnt that in the United States 'Sparrow-hawk' refers to a small *Falco* which hovers like the European Kestrel, a fact which strengthens the point I was trying to make.

cases such a name is insufficiently precise, and it is desirable to add the time and place where the individuals were collected, or the particular laboratory strain to which they belonged, but nevertheless a classification into species is a prerequisite for any accurate communication between biologists.

I have laboured these two points in order to bring out the fact that taxonomists, who identify and name animals and plants, are faced by a contradiction between the practical necessity and the theoretical impossibility of their task. In struggling with this contradiction, they have been led to make important contributions to our knowledge of evolutionary processes. However, as we shall see, their job is not as impossible as the above discussion might suggest. In a wide range of cases clear discontinuities do exist; the division of animals and plants into species is a fact which is often true, and independent of the practical desirability of so classifying them. It is now necessary to describe the situations in which such discontinuities between species are present, and also some cases where they disappear.

It will be helpful to start with a group of animals familiar to most readers. Many people who live in the country or on the outskirts of towns put out food for birds during the winter. In most parts of this country there are at least ten species which are regular visitors to bird-tables during the winter; they are the Blackbird, the Song Thrush, the Great, Blue, and Coal Tits, the House Sparrow, the Starling, the Chaffinch, the Hedge Sparrow, and the Robin. Now any careful observer would quickly learn to recognize these ten kinds of birds; more accurately, he would learn to recognize thirteen different kinds, since in the case of the Blackbird, Chaffinch, and House Sparrow the two sexes are different in the colours of their plumage. It would be difficult, without killing and dissecting a number of specimens, to realize that the brown hen Blackbird is a female of the same species as the black yellow-billed cock, whereas the brown Hedge Sparrows include both males and females, and are not female Robins, which they closely resemble in shape and comportment.

Thus a number of independent observers would soon come to recognize the same thirteen different kinds of birds. They would agree in their classifications, because there are no intermediates between, for example, Coal Tits and Great Tits, although the two species certainly resemble one another more closely than they do Robins or Blackbirds. This is not to say that there are no differences between individual Great Tits or individual Coal Tits; as pointed out earlier, such differences may be great enough to make individual recognition possible. Nevertheless Great Tits and Coal Tits are different categories between which no intermediate or ambiguous forms exist.

If the period of observation were extended into the breeding season, two facts would emerge. First, the hen Blackbirds, Chaffinches, and House Sparrows would be recognized as the females of their respective species, reducing the number of species recognized to ten. Second, it would be noticed that birds of a given species always mated with a member of their own kind. Thus a classification originally based on external appearances would be found to correspond with one based on breeding groups, except for the corrections to be made because the two sexes of a single species had at first been classified in different groups. Similar corrections have had to be made in the history of classification, particularly of invertebrate animals, when it has been recognized that two forms, originally regarded as different species, are in fact only different stages in the life cycle of the same species. For example, it would be impossible without direct observation to tell which kinds of caterpillar develop into which kinds of moth.

There is another situation in which two forms, although sharply distinct in appearance, are nevertheless correctly classified into a single species. For example, some Common Guillemots have a white ring surrounding the eyes, from which a white line extends towards the back of their heads; these are the so-called 'bridled' forms. These have been studied by Southern, who finds that the two forms nest together on the same cliffs, and that mating is at random, there being no tendency for bridled birds to mate with each other rather than with unbridled ones. The situation is therefore one of polymorphism, similar to those discussed earlier in moths and snails. The fact that a sharp distinction remains between bridled and unbridled birds, in spite of random mating, suggests that the difference is due to a single Mendelian factor, which also appears to cause slight differences in skull structure and in the shape of the tail feathers. However, the relevant point here is that even when two forms are distinct, and the difference extends to several features, this is not regarded as a satisfactory reason for placing them in different species, if it is known that they breed together freely in the wild. A corollary, the importance of which will become apparent below, is that it would be difficult or impossible in such cases to decide whether to place such forms in the same or in different species if nothing were known of their breeding behaviour.

We can conclude, then, that for the wild birds of Britain it is usually a simple matter to decide whether two individuals should be placed in the same species by examining their structure and plumage, although in some cases it is necessary to confirm such decisions by observing whether two forms commonly breed together in the wild. The qualifications 'commonly' and 'in the wild' are important. Consider first the qualification 'commonly'. In England the two kinds of gulls, the Herring Gull and the Lesser Black-backed Gull, are, according to the methods outlined above, classified as different species. They are sharply distinct in the colour of their legs and beaks, and in the presence of a dark-grey mantle in the latter species. In breeding behaviour too, the two forms behave as distinct species; Herring Gulls mate with Herring Gulls and Lesser Black-backs with Lesser Black-backs. However, Tinbergen has reported interbreeding between the two species in the wild in Holland, although he emphasizes that it is a rare exception, not the rule, probably occurring mainly when an individual attaches itself to a colony of another species, in which it can find no mate of its own kind. Such rare inter-specific matings give rise to recognizable and fertile hybrids. But such interbreeding is too rare to lead to a merging of the characteristics of the two species, which therefore remain distinct. It is in fact not uncommon to find that two closely related species occasionally interbreed in the wild. Provided that such interbreeding is the exception rather than the rule, the two forms are still regarded as different species.

It is also important to remember that two species which, although living in the same area, do not interbreed in the wild, can nevertheless often be induced to hybridize in captivity, and may produce viable and fertile offspring. The lessons to be learnt from such hybrids will be discussed in a later chapter. If crossing occurs in captivity but not in the wild, it cannot have the effect of blurring the distinction between the two wild populations, which are therefore still regarded as distinct species.

For the birds of a given geographical region, for example Great Britain, the picture which has emerged is one of a series of distinct and easily recognizable species, between which few or no intermediates exist. Birds breed with members of their own species, and hybridization in the wild is rare or absent. In consequence each species can become adapted to a particular mode of life. A division of labour in the exploitation of the environment is in animals based on inherited differences between reproductively isolated populations. This, however, is not a sufficient explanation of the way in which specific distinctions have originated. In later chapters we shall have to consider how natural selection has acted to produce the discontinuities between species. For the present it is enough to recognize, first, that it is the free interbreeding within species and the absence of hybridization between them which are responsible for the relative uniformity of structure of the members of a given species and for the absence of intermediate forms, and second, that direct observation of the breeding habits of animals and plants is a necessary method of confirming classifications based on similarities of structure.

In practice, however, the classification of the animals and plants of a given region into species is seldom as easy as it is in the case of birds. In Chapter 10 some observations on natural selection in wild populations of the fruitfly *Drosophila pseudo-obscura* were described. When this species was first kept in captivity by Dobzhansky and his colleagues, two 'races' were recognized, A and B. Flies of one race mate more readily in captivity with flies of their own race than of the other. Hybrids can, however, be obtained in captivity, the female hybrids being fertile but the males infertile. The two 'races' exist in the same areas over a wide range in the United States. Fortunately there are structural differences between the chromosomes of the two races, which can be recognized in the first generation hybrids and in later generations as 'inversion loops' (see page 129). A study of chromosome structure in wild populations has shown that in fact hybridization must be rare in the wild, and exchange of genetic material between the two races extremely rare or absent. Therefore the two races A and B are now considered to be separate species, the second species being named, for obvious reasons, *Drosophila persimilis*. It has been found that there are differences between the genitalia of the males of the two species, too slight to have been noticed until the above observations on breeding behaviour provoked a deliberate search for them, but sufficient to make it possible to identify a single specimen. No such difference has yet been found by which the females of the two species can be distinguished. Thus a female of *D. persimilis* must be defined as one which will mate readily with a male of that species, and produce fertile offspring of both sexes. Such pairs or groups of similar species are sometimes referred to as 'sibling' species; many other examples could be given from the genus *Drosophila* alone.

A still more remarkable situation has been described by Sonneborn and his colleagues in the slipper animalcule, *Paramecium aurelia*. It has already been explained (page 206) how, although it is impossible to recognize males and females in these protozoa, it is possible to distinguish two 'mating types'. Every individual belongs to one or other mating type, and will conjugate only with members of the other mating type. It has, however, been found that the 'species' *P. aurelia* consists of a number of 'varieties'; conjugation between members of different varieties rarely takes place, and if it does the partners die after conjugation. Thus it would be more in line with the methods of classification adopted in metazoan animals to regard each of these varieties as a separate species. A similar situation exists in a number of other ciliated protozoa.

Both *Drosophila* and *Paramecium* have been subjected to intensive study by geneticists. The consequences of this study have been, not to invalidate the concept of reproductively isolated species, but to show that a classification based on differences in structure may be insufficient to detect all the specific distinctions which in fact exist. Probably similar difficulties will emerge when other groups have been studied in equivalent detail.

It is therefore worth asking why it is that in birds a classification based on visible structures has agreed so well with one based on breeding behaviour. The answer lies in the methods by which birds are able to recognize members of their own species. Clearly some method of species recognition is necessary during the breeding season if individuals are to mate with members of their own species. The method of recognition in any group will depend on which sense organs are particularly well developed in that group. Birds probably rely little on their sense of smell, but have well-developed colour vision and an acute sense of hearing. Thus their sensory equipment is similar to our own, and we are therefore well fitted to observe differences in plumage colour, display movements, and song. We find specific differences in birds relatively easy to recognize because we notice the same features which the birds themselves use during the breeding season. It is interesting that the three sibling species of leaf warblers in Britain, the Chiffchaff, Willow Warbler, and Wood Warbler, were not at first distinguished by their plumage, which is strikingly similar in the three species, but by their songs, which, as Gilbert White was the first to appreciate, are wholly dissimilar.

In many other animal groups there is a far greater development of the chemical sense of smell, and species recognition probably depends on this sense. In the case of *Paramecium*, for example, it is difficult to see how the readiness of one individual to conjugate with another could be determined, save by the detection of chemical differences between varieties and between mating types. The importance of the chemical sense is not, however, confined to primitive organisms; in most mammals, other than Primates, it is more important than either sight or hearing. Now, for reasons which will be discussed later, natural selection tends to exaggerate those differences between species which are important in specific recognition. In birds, in which we can directly perceive these differences, classification is easy; in animals relying on a sense of smell, it is often difficult.

So far I have discussed classification into species only for populations which satisfy three requirements, namely that they reproduce sexually, that they inhabit the same geographical region, and that they live in the same geological period. All these requirements must be met if interbreeding in the wild is to be used as a criterion of membership of a species. We must now consider how the validity of specific distinctions tends to breakdown if any one of these requirements is not satisfied.

Little need be said of the requirement that the populations must be contemporaneous. If the characters of a population evolve gradually with time, this process will in the end result in a difference between the initial and final populations as great as that between two contemporary species. It is, however, not possible to say whether the members of the two populations would have interbred had they met, and if so whether they could have produced fertile offspring. All the same it is convenient for palaeontologists to use the same system of naming as do other biologists; hence the only reasonable course is to give two fossils or groups of fossils different specific names if the difference between them, is of the same order of magnitude as that between contemporary species of the same group. In practice it is rare to find a series of sedimentary deposits, each containing populations of fossils descended from populations in earlier deposits, over a sufficient period of time for the gradual transformation of species to be observed. More often, slight changes in the conditions in which the sediments were laid down cause the replacement of one fauna of fossil species by another, probably not descended from the earlier forms but entering the area from elsewhere. In the same way the present fauna of southern England is not in the main descended from the fauna which inhabited the same area during the ice ages. Rather, members of the original fauna have become extinct, or have migrated northwards, being replaced by immigrating

species from farther south. Consequently the boundaries between fossil species are usually fixed by more or less accidental
discontinuities in the fossil record; they do not necessarily reflect
actual discontinuities in the evolution of species in time.

Similar difficulties arise in animals or plants which reproduce
asexually – either vegetatively or by some form of parthenogenesis. Parthenogenesis is not uncommon in animals, but
in most cases it is a cyclical process, alternating with sexually
reproducing generations; this is the case, for example, in aphids
(e.g. greenfly). Such cyclical parthenogenesis does not alter the
validity of the species concept. Permanent parthenogenesis is
rare in animals, occurring sporadically throughout the animal
kingdom. In the brine shrimp, *Artemia salina*, there is a sexually
reproducing 'race', and a number of parthenogenetic races,
most of them having a larger number of chromosomes than the
sexual race, and probably evolved from it in the fairly recent
past. Similar parthenogenetic races are known of species of
wood lice (*Trichoniscus*), water fleas (*Daphnia*), and moths
(*Solenobia*). The fact that in all these cases the parthenogenetic
races closely resemble sexually reproducing forms suggests that
they are of recent origin, and hence that this method of
reproduction has disadvantages which in the long run prevent
such races from undergoing an adaptive radiation and so giving
rise to larger taxonomic groups all of whose members reproduce
parthenogenetically.

In plants, both vegetative reproduction and self-fertilization
are relatively common. For example, asexual reproduction is
the rule in dandelions (*Taraxacum* spp.), although they probably
originated from sexually produced species hybrids. In the
absence of interbreeding, which could produce some degree of
uniformity within species together with sharp discontinuities
between them, a bewildering variety of distinct asexually
reproducing lines of dandelions can be recognized. To give a
different specific name to each such variety would be impractical, and for convenience Turrill has suggested naming
only a few of the more distinctive varieties. A similar situation
exists in other groups, for example, blackberries (*Rubus* spp.)
and roses (*Rosa* spp.); in all such cases classification into species

and the giving of specific names is retained for convenience, but no longer corresponds to any real features of the patterns of variation of the plants in nature.

The major difficulty, however, in the classification of animals and plants into species arises when studying populations from different geographical areas. As was emphasized in Chapter 8, species with a wide range of distribution in space tend to be divided into a series of populations which differ from one another to a greater or less extent. Such variation may be continuous, the populations at the boundaries of the distribution being connected by a series or 'cline' of intermediate forms, or discontinuous, the species being broken up into a series of more or less well-defined geographical races. In the early days of taxonomy, it was often the practice to give a new specific name to a specimen or group of specimens collected from a new area if they differed sufficiently from already known forms for the identification of individual specimens to be possible. But this practice has been in the main abandoned, partly because further study has often revealed the existence of a series of forms intermediate both in their place of origin and their structure, and partly because, even where such intermediates do not exist, the number of distinguishable local populations may be so great as to render it absurd to erect a separate specific category for each of them. The present practice, therefore, where a series of races replace one another geographically, is to include them all in a single species, and to regard the local races as 'subspecies'. Thus, for example, the Lesser Black-backed Gulls of Britain and of Scandinavia differ slightly but consistently in the colour of their legs and mantle, and accordingly are regarded as different subspecies, *Larus fuscus graellsii* and *Larus fuscus fuscus*, respectively, of the same species *L. fuscus*.

There must, however, be some degree of difference between two geographically isolated populations which will justify placing them in different species. Since all degrees of difference between such populations may exist, from barely recognizable statistical differences to clear-cut differences of the same order of magnitude as separate different species inhabiting the same area, the decision in any particular case is to some extent an arbitrary one, although it is usual to erect two specific categories only when the difference is of the latter kind.

This difficulty arises because, with geographically isolated populations, it is impossible to apply the test, 'Do the two forms interbreed freely in the wild?' However, although the decision to place two such populations in different species is often a matter of individual judgement, it does nevertheless imply a hypothesis about the future, namely that the two forms will remain distinct, even though their ranges of distribution may come to overlap, and so will continue to diverge in structure and habits. In contrast, the grouping of two species into the same genus, as for example the Herring Gull *Larus argentatus* and the Lesser Black-backed Gull *L. fuscus* into the same genus, *Larus*, is a hypothesis about the past, since it implies that a common ancestor of these two forms existed in the more recent past than did the common ancestors of either of these gulls and of other related genera, for example, the Terns (*Sterna* spp.) or the Kittiwake (*Rissa*). This is not to say that all the hypotheses, either about the future or the past, implied by our classification are necessarily correct, although those implied by the names *Larus argentatus* and *L. fuscus* probably are so.

Applying this method of classification to the human populations of the world, it is clear that all human races should be regarded as members of the same species, although some might warrant sub-specific rank. Wherever migrations have brought together two such races in the same country, intermarriage has taken place, although social pressures and prejudices may have slowed down the final mixing of the two populations. It seems very likely that the future will see a continued mixing of human races, and blurring of distinctions between them. In the same way, although with less confidence, it was suggested in Chapter 8 that the Hooded and Carrion Crows are best regarded as subspecies of a single species, since so long as an area of free interbreeding between the two forms exists, it seems unlikely that they will continue to diverge genetically.

Typically, no opportunity arises for populations from the two ends of the geographical range of a species to meet in the wild. But an illuminating exception occurs in the case of so-called 'ring species', of which two examples from British birds can be given. The first concerns two species of gulls already discussed, *Larus argentatus* and *L. fuscus*, which, as described above, behave in England as distinct species, rarely interbreeding, and differing in their choice of nesting sites and in the areas to which they disperse during the winter. It was shown by Stegmann that there is in fact a chain of ten recognizable forms or subspecies of gulls belonging to the *fuscus-argentatus* group, differing mainly in the colour of their legs and mantles, forming a ring round the North Pole, from the British Isles through Scandinavia, northern Russia, and Siberia to the Behring Straits, and thence via Alaska to Canada and back to Britain. The two forms found in Britain are the terminal links in this chain. Since their breeding behaviour demands that they be placed in different species, it is necessary that the boundary between *L. fuscus* and *L. argentatus* be drawn between two intermediate populations in the chain. Fisher and Lockley have suggested that the boundary be drawn between the dark-mantled form from central Siberia and a paler form from the north-west of Siberia. But they emphasize that this choice is an arbitrary one, and probably does not correspond to the region inhabited by the ancestral gull population which gave rise to the whole chain. The situation is further complicated by the existence of two 'side-chains' of subspecies, one spreading from Lake Baikal through the Caspian and Black Seas to the Mediterranean, and a second northwards from Canada to Greenland, where there is a very pale form originally classified as a third distinct species, *L. glaucoides*.

Thus three populations of gulls from the north Atlantic, originally classified as three distinct species, *Larus argentatus*, *L. fuscus*, and *L. glaucoides*, the two former being still so classified because they remain distinct in the area of overlap in Britain and north-west Europe, have since been found to be connected by a complex series of intermediate forms. How can we explain this situation? In the absence of direct evidence we can only guess, but its origin may have been as follows. An ancestral gull population, possibly breeding in the Behring Straits, increased in numbers and colonized new areas both to the east and to the west. Although gulls have great potential powers of dispersal, in practice individuals usually return year after year to breed in the same area, often in the same colony, as that in which they were reared. Consequently a fair degree of reproductive isolation would exist between the original population and the new colonies derived from it. This partial isolation would make possible some divergence in structure and habits. Finally, populations spreading eastwards must have met similar populations which had spread to the west, the meeting-place being on the other side of the globe in the north Atlantic. When this meeting took place, the two forms were sufficiently different to remain distinct.

The British Great Tit, *Parus major*, forms part of a similar though less complex ring species, first described by Rensch. A series of intergrading forms of this species extends from Britain eastwards through Europe, Persia, and India to Malaya, and thence northwards through China to Japan. A second series of forms extends eastwards from Europe, north of the desert regions of southern Russia and of the Himalayas to Mongolia and northern China. This second series is probably of fairly recent origin, the area being colonized only after the end of the ice ages. In northern China the terminal links in these two chains of forms overlap in the Amur valley, where they do not inter-breed. In this case the 'ring' has been formed round the desert and mountain regions of central Asia, but as in the case of the circumpolar ring of gull species, the terminal forms behave as distinct species.

These two examples show how arbitrary must be any attempt to classify animals and plants from different geographical areas into species. In the case of two species inhabiting the same area, interbreeding within a species and its absence between species are at the same time the causes and the criteria of specific distinctness. In the absence of such criteria, owing either to asexual reproduction or to isolation in space or in time, the boundaries between species lose their sharpness, and classification into species is retained more because of convenience than because it reflects reality.

The example of ring species also suggests that geographical

isolation may itself be an important cause of divergence between populations, and hence of speciation. Ring species of this kind, however, are the exception rather than the rule. In the next chapter I shall discuss other lines of evidence which suggest that isolation, geographical or otherwise, has been an important stage in the process whereby single populations have become separated into different species.

CHAPTER 14

The Origins of Species

In 1835 Charles Darwin, while naturalist on board H.M.S. *Beagle*, visited the Galapagos Islands. Later he wrote in his *Evolutionary Notebook*, 'Had been greatly struck ... on character of S. American fossils – and species on Galapagos Archipelago. – These facts origin (especially latter) of all my views.' In 1854 he wrote of the finches of the Galapagos, 'seeing this gradation and diversity of structure in one small, intimately related group of birds, one might really fancy that from an original paucity of birds on this archipelago, one species had been taken and modified for different ends'. This was the first public statement by Darwin of his evolutionary views.

In fact a study of the structure and distribution of these finches has not only confirmed Darwin's view that they are indeed the modified descendants of a single species, but also indicates the ways in which speciation took place. The picture which emerges is probably of wide application. The account which follows is based on the work of Lack, who visited the archipelago in 1938 with the express purpose of studying Darwin's finches.

The Galapagos are a group of volcanic islands lying on the equator in the eastern Pacific, 600 miles from Ecuador. The isolated island of Cocos lies 600 miles to the north-east; to the west there is no land for 3,000 miles.

The largest island, Albemarle, is some 80 miles long and rises to a height of 4,000 feet. There are several islands rising to 2,000–3,000 feet, and a number of small low-lying islands. Three general types of habitat are available; the coastal plain is arid, with thorn bushes, cactus, and prickly pear, and some

231

open ground free of vegetation where lava flows have occurred recently. On higher ground, with higher rainfall, there is humid forest; such forests occur only on the larger islands, James, Indefatigable, Albemarle, and Charles. On the highest ground the forest gives way to more open country, but since this region has not been colonized by finches it does not here concern us.

Darwin's finches have been divided into fourteen species, of which one occurs only on the distant island of Cocos, and the other thirteen on the Galapagos. All are greyish-brown, short-tailed birds, although in some species the males are black, and they resemble one another closely in their courtship displays, in their nests and eggs, and in many features of internal anatomy. They have accordingly been grouped together in a subfamily, the Geospizinae, of the finches. The most striking differences between the species are in size, in the shape of the beak and, associated with this, in feeding habits. In describing the various species I shall use the English names suggested by Lack, since these are more easily remembered than the Latin names. Of the thirteen species on the Galapagos, six are ground finches, feeding on seeds, or in one case on cactus, and having beaks resembling those of the more typical seed-eating finches. Of these six species, all except one, the Sharp-beaked Ground Finch, are confined to the arid coastal regions. Six further species are tree finches, inhabiting humid forest and feeding on insects. Perhaps the most remarkable of these is the Woodpecker Finch. This bird resembles the woodpecker in its ability to climb up and down vertical tree trunks, and in its habit of excavating holes in branches in search of insects. But whereas a woodpecker inserts its long tongue into cracks to capture insects, the finch picks up a twig or cactus spine, which it pokes into the crack, subsequently dropping the twig in order to seize any insect which emerges. This is an unusual example of the use of a tool by an animal other than man; it will be most interesting to discover how far the development of the habit in individuals is dependent on copying other birds. Of the other five species of tree finches, three feed on insects in the upland forests, one is vegetarian, and there is one insectivorous species inhabiting the coastal mangrove swamps on Albemarle and Narborough.

Finally there is a Slender-billed Warbler Finch present on all the main islands, a species which closely resembles true warblers in its form and habits.

It seems certain that at some time in the past a flock of fairly typical seed-eating finches reached the Galapagos Archipelago. At the present time there are very few other species of passerine birds on the islands. All are insectivorous, including a Warbler, a Martin, two species of Flycatchers, and several related species of Mocking Birds. In all cases these species closely resemble species from the American mainland, and it is therefore very probable that they reached the islands at a later date than did the finches. In any case, they may compete for food with the Warbler Finch and perhaps with the insectivorous tree finches, but not with the seed-eating, cactus-eating, or wood-boring finches. Thus on their arrival, and to a large extent at the present day, the finches have met with little competition from other passerine birds. In these circumstances they have undergone an 'adaptive radiation', adopting ways of life normal to woodpeckers, tits, and warblers as well as those normal to finches.

Such adaptive radiation in the absence of competition is a common feature of evolution. Thus in Australia the marsupial mammals, which are now extinct on the mainland of Asia, have evolved carnivorous types, the Tasmanian wolf, Tasmanian devil, and native cat, as well as a variety of herbivorous forms, such as the leaping kangaroos, the burrowing wombats, the climbing phalangers and koala bears, and even the gliding phalanger, *Petaurus*. Such a radiation has been possible because of the absence of placental mammals (other than bats and small rodents) in Australia. In South America the marsupial mammals also underwent a considerable adaptive radiation, but mainly of carnivorous and insectivorous types; the opossums are surviving examples. Here the evolution of herbivorous marsupials was probably prevented by the presence of an endemic group of herbivorous placental mammals, now mainly extinct, but which in their time underwent an adaptive radiation paralleling that of the herbivorous mammals of the rest of the world. In New Zealand, where bats are the only endemic mammals, there evolved a number of kinds of large flightless birds. On Madagascar, which has been colonized by relatively few mammalian groups, the lemurs, which on the mainland have largely been replaced by their descendants the monkeys, have evolved carnivorous and insectivorous as well as fruit-eating species.

Here I must introduce an important idea originally due to Gause. It can be stated as follows: two or more similar species will not be found inhabiting the same locality unless they differ in their ecological requirements, for example in their food or their breeding habits, in their predators or their diseases. This conclusion follows from the argument that, if two species were identical in such respects, it is certain that there would be a difference, however slight, in their efficiency, and so the less efficient species would become extinct in face of competition from the fitter one. So formulated, the statement cannot rigidly be proved or disproved, but the idea behind it is important for two reasons. First, it provides an impetus to ecologists confronted by two or more species inhabiting the same locality to seek for the ecological differences between them. Second, it provides an explanation for a type of distribution commonly found, of which an example afforded by the Sharp-beaked Ground Finch will be described below.

Where two or more similar species are found within a given region, they may occur in different local habitats within that region. For example, of the three English pipits, the Meadow Pipit is a bird of open moorland, the Tree Pipit of open woodland, and the Rock Pipit is confined to rocky shores. Localities inhabited by these three pipits may border on one another but seldom overlap. Alternatively, similar species may be found in the same habitat, but differ in the uses they make of it. Thus Hartley has shown that although Great, Blue, Coal, and Marsh Tits may all be common in a single wood, they tend to feed at different heights above the ground and at different distances from the main trunk.

These ideas will help to explain some features of the distribution of Darwin's finches. Three of the ground finches, the Large, Medium, and Small Ground Finches, occur together in the coastal regions of the main islands. Lack found that although the three species often feed on the same things, the large hard manzanilla fruits were taken only by the two larger species, while the Small Ground Finch fed predominantly on smaller seeds, particularly of grasses. While there is a considerable overlap between the foods taken by the three species, it may be that there are sufficient differences to account for their survival side by side in the same localities.

The distribution of the Sharp-beaked Ground Finch is of particular interest. It is the only species of ground finch which breeds in the humid forests. On the central islands it is confined to this zone, but on Culpepper, Wenman, and Tower, on which the Small Ground Finch does not occur, it breeds in the arid zone. This suggests that the Sharp-beaked Ground Finch, although perfectly capable of surviving in the arid zone in the absence of competition from the Small Ground Finch, cannot do so if the latter species is present.

The division of the finches into a number of species has enabled them more effectively to exploit the environment, by becoming specialists in using the various kinds of food available. The next problem to be discussed is how the original immigrant population broke up into a series of species which now do not interbreed even where they inhabit the same islands.

It has been emphasized in earlier chapters that the tendency for a species to be split up into a series of differentiated and locally adapted populations is often counterbalanced by migration and interbreeding between such populations, which results in a levelling out of the differences between them. However, on an archipelago different populations are isolated from one another by the intervening sea. It is, of course, true that the distances between different islands in the group are such that the finches are physically capable of flying from island to island. Such migration must have taken place, but it is probably a rare and accidental event. In such cases, then, island populations may be effectively isolated from one another for considerable periods, with little or no interbreeding, giving ample time for the evolution of divergence in habits and structure. From time to time, however, a flock of birds from one island may migrate to another. Should the descendants of such a flock become established in their new home, and should the immigrant and resident populations be sufficiently different for interbreeding between them to be rare or absent, then the two populations can only be regarded as different species. The various ways in which differences may lead to a barrier to interbreeding will be discussed in later chapters. For the present, we must see how far the distribution of the Galapagos finches confirms the idea that the origin of the differences between species lay in the isolation of island populations.

The description which was given above of the various kinds of finches may have suggested that all the members of a given species are alike, and sharply differentiated from the other species. This is very far from the truth. For example, the Warbler Finches from different islands vary in the colour of their plumage and the size of their beaks and wings, and there are similar differences between island forms of the Sharp-beaked Ground Finch. Each of these species has therefore been divided by taxonomists into a series of subspecies. In most cases in which two forms have been classified as different species, they occur together on the same island in at least part of their range without interbreeding, as do, for example, the Small, Medium, and Large Ground Finches. In such a case the erection of three distinct species is confirmed by the failure to interbreed. However, in the case of the two cactus finches, *Geospiza scandens* and *G. conirostris*, the two forms occur on different islands, replacing one another geographically as do the different subspecies of the Warbler Finch. The decision to recognize two distinct species of cactus finch depends on the degree of anatomical difference between them, which is rather greater than the differences between the various island forms of Warbler Finches or of Sharp-beaked Ground Finches. Thus all degrees of difference exist between island populations, from those which are barely recognizable to those which have been considered great enough to justify specific rank. Thus the isolation between island populations has resulted in a divergence of structure whose extent depends on the length of time for which the populations have been isolated and on the difference in conditions, and hence of selection pressures, on different islands.

The importance of isolation in the divergence of Darwin's finches can be illustrated in another way. The table below, modified slightly from that given by Lack, groups the islands according to their degree of isolation, and lists the total number of species present, and the number of endemic subspecies, i.e. of subspecies found only on that island.

Degree of Isolation	Island	No. of Resident Species	Endemic Subspecies not found on other Islands	
			Number	%
Very extreme	Cocos	1	1	100
Extreme	Culpepper and Wenman	4	3	75
Marked	Hood	3	2	67
	Tower	4	2	50
	Chatham	7	2	29
Moderate	Abingdon and Bindloe	9	3	
	Charles	8	2	25
	Albemarles and Narborough	10	2	20
	Barrington	7	1	14
Slight	James	10	–	0
	Jervis	9	–	0
	Indefatigable	10	–	0
	Duncan	9	–	0

As this table shows, the greater the degree of isolation of an island, the smaller the number of species which inhabit it, but the greater the proportion of subspecies peculiar to it. There is a smaller number of species on the outlying islands because there is little possibility of a species inhabiting such an island evolving *in situ* into two distinct species, and there is less chance of colonization by species which have evolved elsewhere. The greater proportion of endemic subspecies on such islands shows that those few species which have colonized them have subsequently evolved their own special characteristics.

This conclusion is confirmed by a study of birds from other areas. The table below is a simplified version of one given by Mayr for birds from Manchuria, where few barriers to migration exist, and from the Solomon Islands.

Frequencies of different kinds of Species of Birds in
Manchuria and in the Solomon Islands, after Mayr

	Manchuria 107 spp.	Solomon Is., 50 spp.
Species with restricted range, not divided into subspecies	2%	18%
Species with wide range, not divided into subspecies	14%	2%
Species divided into subspecies	83%	46%
Superspecies†	1%	34%

† i.e. groups divided into a number of populations, some of which are sufficiently different to be regarded as distinct species; e.g. the *Larus argentatus-fuscus* group described in the last chapter.

Thus the barriers to interbreeding which exist in the Solomons have favoured the evolution of subspecific differences; of species with a wide range, only one species out of 50 (2 per cent) on the Solomons is not so divided, compared to 15 out of 107 (14 per cent) in Manchuria. Such barriers have also favoured the evolution of still wider differences between local populations, as is shown by the large proportion of 'super-species' in the Solomons.

From studies of this kind the idea has become accepted particularly through the influence of Rensch, Mayr, and Julian Huxley, that geographical isolation is often the first stage in the splitting of a single species into two. This view was reached largely by a study of variation in birds. There is, however, some evidence that it is also true for other animal and plant groups, although the actual physical barriers to migration may vary. For example, in fresh-water fishes the populations in different river systems are effectively isolated from one another, and often differ from one another genetically. Studies on the land snails of the Pacific islands have shown that the populations in different wooded valleys are partly isolated from one another by intervening rocky ridges, although occasional dispersion from one valley to another must take place. Several species of these snails are polymorphic for size, colour, and direction of coiling of the shell. The frequency of different types varies from valley to valley, and particular forms may be wholly absent in some populations. This suggests that the characteristics of the snails in a particular valley may be partly a matter of chance, depending on the characters of the relatively few individuals originally entering the area, and whose descendants have populated it. The importance of such 'founders' of local populations will be discussed further below (page 275).

Since Britain is itself a large island, with a series of smaller islands off its coasts, we would expect to find, in those groups in which transport across the sea is a rare and accidental event, that the British population differs from that on the mainland of Europe, and that distinctive populations occur on many of our smaller islands. This is in fact the case, for example, for our smaller mammals, including the long-tailed field mouse *Apodemus*, of which there are subspecies on St Kilda, the Hebrides, and Shetland, for the bank vole *Clethrionomys*, and for the short-tailed vole *Microtus*. It is a curious fact that the bank voles from Skomer, off the Pembrokeshire coast, differ from the mainland forms not only in their greater size and in colour and skull structure, but also in their remarkable tameness.

So far we have looked for the origin of reproductive isolation between two populations in the form of a geographical barrier, the nature of the barrier necessary to produce a given degree of isolation depending on the habits and powers of dispersal of the species in question. We must now turn to some facts which suggest that reproductive isolation between two populations may evolve while both are inhabiting the same geographical area. It is quite common to find two or more 'biological races' of a given species in the same locality, differing little or not at all in appearance, but nevertheless quite distinct in their food preferences. Some examples will be given of this phenomenon for insects feeding on apples and other fruits.

The caterpillars of the moth *Hyponomeuta padella* feed on apple and on hawthorn trees. There is in this species considerable variation in the colour of the fore-wings of the adults, from dark grey to white, but although the dark forms are commoner among moths developing from caterpillars from hawthorn, and the white forms commoner in those from apples, the difference is only one of the relative frequencies of the different colour varieties, and the adults are otherwise indistinguishable in appearance. However, the adult moths can usually be distinguished by their egg-laying preferences, since about 80 per cent of individuals reared on apple lay their eggs on apple trees, and about 90 per cent of those from hawthorn lay their eggs on hawthorn. There is also a tendency for moths to mate with partners raised on the same food plant; in experimental conditions such 'assortative' matings were about twice as common as were matings between individuals raised on different plants. Finally, the caterpillars show a strong preference for the food plant on which their mothers were raised, although they can be induced by starvation to feed on the 'wrong' food plant; in such cases the resulting adults are often infertile. In this case, then, the two races are not completely isolated since some interbreeding probably takes place in the wild; they are best regarded as subspecies of a single species.

A slightly greater degree of divergence has occurred between two forms of the American two-winged fly *Rhagoletis pomonella*. The larvae are of two sizes, the larger attacking apples and the smaller blueberries and huckleberries. No structural difference other than size has been detected. It is very difficult to persuade adults of these two forms to cross, and equally difficult to raise larvae of one form on the food plant proper to the other. These two kinds of flies probably never interbreed in the wild, and should be regarded as distinct species, although no difference other than size can be recognized.

A third example can be given in which the degree of divergence is still greater. Two forms of the bug *Psylla mali* feed respectively on apples and on hawthorn. The adults differ only in size, but slight morphological differences have been found between the nymphs. It has proved impossible to get these two races to cross in captivity, or to persuade either to lay eggs on the other's food plant.

Now although these three insects are only distantly related to each other, belonging to different orders, the facts given above suggest that all are undergoing a similar process of differentiation into two distinct species, a process which is still in its early stages in *Hyponomeuta* and which is effectively completed in *Psylla*. Further, since in each case the two forms are found inhabiting the same areas, the facts suggest that the process of divergence can proceed in the absence of geographical isolation. This raises a difficulty. It is easy enough to imagine that individuals would arise in a population, through the chance processes of mutation and segregation, genetically better adapted than the average to cope with a new food plant. But unless such individuals tend to lay their eggs on that plant, and to mate with individuals having similar adaptations, the adaptation will soon be lost as a result of random breeding in the population. It is demanding a miracle to suggest the chance origin of a new genotype which simultaneously influences the capacity of the larvae to grow on a new food plant, and the egg-laying habits and mating preferences of the adult. The answer to this difficulty may be contained in the work of Harrison and of Thorpe.

Harrison studied the sawfly *Pontania salicis*, the larvae of which produce galls on willows. A number of biological races of this species exist, forming galls on different species of willow. A race normally forming galls on *Salix andersoniana* was confined to another species of willow, *S. rubra*. In the first generation the majority of the larvae died, but this heavy mortality fell off in later generations. After four generations during which the population was confined to the new species of willow, three further generations were raised in which a choice of species was provided, and it was found that the sawflies continued to lay eggs on their new host plant. It seems very unlikely that four generations of selection for a capacity on the part of the larvae to survive on *S. rubra* could alter the genetic determination of the egg-laying preferences of the adults; it is much more probable that adult sawflies tend to lay eggs on the kind of food plant on which they themselves have been raised.

This conclusion is strengthened by some observations by Thorpe on the ichneumon fly, *Nemeritis canescens*; the ichneumons are insects related to wasps and bees, in which the winged adult females lay their eggs in the bodies of other insects, within and at the expense of which the larvae develop. *N. canescens* normally parasitizes caterpillars of the meal-moth *Ephestia*. Thorpe reared larvae in the caterpillars of another moth, *Achroia*, and observed the behaviour of the adult females so obtained. Female ichneumons hunt for their prey by means of an acute sense of smell. Thorpe found that all *Nemeritis* females were attracted by the smell of their normal host, *Ephestia*; this preference is presumably genetically determined. However, *Nemeritis* females which had been reared in *Achroia* caterpillars, or which had been in close contact with *Achroia* immediately after emerging as adults, were also attracted by the smell of *Achroia*. Thus it is also genetically determined that a female *Nemeritis* shall be attracted by the smell of the insect species in which she developed as a parasite. Normally these two processes would reinforce one another in determining the attraction of females to *Ephestia*. But the attraction of females to unusual hosts, if they themselves were reared in such hosts, could play a part in the evolution of a biological race adapted to a new host species. In such a case, the egg-laying preferences of females would be transmitted from generation to generation in the same way as languages are transmitted in our own species; it is genetically determined that human beings can learn to talk, but not that they shall learn English rather than French, or vice versa.

It is now possible to see how a species may split into biological races without geographical isolation, and without demanding that a number of different, genetically determined adaptations should arise simultaneously and by chance. If for any reason a female of, for example, a plant-eating species lays her eggs on an unusual plant, two things will follow. First, the larvae will be exposed to new conditions, and intense selection of genotypes adapted to those conditions is likely; the high mortality which Harrison observed in *Pontania* reared on a new species of willow confirms this. Second, the females which develop on the new plant will tend to lay eggs on that plant, not because they differ genetically from the rest of the species, but because they have been conditioned during larval life. Further, in species which mate soon after emergence, there is a fair probability of a female which has developed on the new host plant mating with a male from the same plant or group of plants, simply because she is more likely to meet such a male. In this way a population of insects can arise which, although not completely isolated reproductively from the rest of the species, may yet be sufficiently isolated by its habits to diverge genetically from the rest of the population. At first the main genetic change to occur in such a population will be that resulting from the intense natural selection for the capacity of the larvae to survive on the new host plant. However, any genetic changes which reinforce either the tendency to lay eggs on the new plant, or to mate with members of the newly adapted population, will be favoured by selection. Thus in *Hyponomeuta* there is already a tendency towards assortative mating, which in *Psylla* is absolute, while in *Nemeritis* there is a genetically determined preference for the normal host *Ephestia*, although it is probable that once *Nemeritis* females laid eggs in *Ephestia* for the same reason that they can today be induced to lay eggs in other hosts – because they themselves developed there.

Such a process of differentiation into biological races is a probable consequence whenever a plant-eating species lays eggs on two or more kinds of food plant, or a parasitic species lays its eggs in several host species. Divergence will clearly be favoured in those species which mate on or near the food plant, or, in parasitic species, which mate immediately after emerging from the host, since this will increase the degree of reproductive isolation between populations exploiting different foods. It is difficult to say how far similar processes have been important in speciation in vertebrates. In birds, for example, the choice of breeding territories and nesting sites is influenced by the nature of the territory and the nest in which the individual was raised, and both song and recognition of other members of the species are influenced by experience. It is therefore quite possible that races of birds within the same geographical area might be kept effectively separate by their choice of different habitats. To give an extreme example, domestic pigeons are descended from rock pigeons, *Columba livia*, and London's pigeons are in turn descended from escaped or released domestic pigeons. Yet the London pigeons remain effectively isolated from their wild ancestors by their choice of buildings instead of cliffs as nesting sites. Given time, they might well evolve as a distinct species.

This chapter has discussed processes of speciation which depend on the gradual accumulation of genetic differences between two populations. There is, however, one process which can lead to the immediate establishment of a new species, reproductively isolated from its ancestors. This is the process of hybridization followed by a doubling of chromosome number, or 'allopolyploidy'; it will be discussed on page 267, after something has been said of the reasons for hybrid infertility.

CHAPTER 15

What Keeps Species Distinct?

It has been the argument of the last two chapters that the division of a single population into two parts, which subsequently evolve into distinct species, is possible only if there is some barrier to interbreeding. In the early stages of divergence it seems likely that geographical isolation is the commonest type of barrier, although it was suggested that, in the case of 'biological races', environmentally caused differences in physiology and behaviour can favour the evolutionary divergence of two populations inhabiting the same area.

Several examples were given of populations which, although they probably first diverged when spatially isolated from one another, have subsequently expanded their ranges, so that today they live side by side in the same regions without losing their separate identities. In this chapter I shall describe some of the processes whereby two related populations inhabiting the same geographical region can remain distinct. Such processes have been called by Dobzhansky 'isolating mechanisms'. The essential feature of such mechanisms is that they shall prevent, or greatly reduce, the exchange of genetic material between two populations; in the absence of such mechanisms the two populations will merge into a single interbreeding unit, as appears to be happening at the present time to our own species.

One of the chief methods of studying such isolating mechanisms is to hybridize different varieties and species. In this chapter I shall review briefly the results of such studies in so far as they throw light on mechanisms of isolation. Chapter 16 will describe the genetics of species hybrids, to see whether the differences between species are determined by the same kinds of

245

genetic mechanisms as are differences within species, and to trace the genetic causes of the isolating mechanisms already described.

It is convenient to start with a classification of isolating mechanisms; such a classification has been suggested by Dobzhansky, whose ideas are accepted with slight modifications in the following scheme:

I. GEOGRAPHICAL OR SPATIAL ISOLATION
    Such isolation has been discussed in the two preceding chapters.

II. BARRIERS PREVENTING THE FORMATION OF HYBRIDS
    A. Isolation by habitat.
    B. Seasonal isolation.
    C. Lack of mutual attraction between males and females of different species of animals.
    D. Mechanical isolation – physical non-correspondence between the genitalia or floral parts.
    E. Prevention of fertilization.

III. HYBRID INVIABILITY
    First generation hybrids are formed, but either do not survive at all or survive poorly in competition with the parental types.

IV. HYBRID INFERTILITY
    First generation hybrids fail to produce functional sex cells.

V. HYBRID BREAKDOWN
    First generation hybrids viable and fertile, but later generations weakly or infertile.

Any one of these processes acting by itself could be sufficient to prevent the merging of two populations. Usually, however, several are found to be operating in any given case, although one may be of major importance in the wild. There are, however, a few cases known in which two species, which in nature are found in different geographical areas, are not separated by any other mechanism if they are brought together in the same place. For example, if the Blue-winged Teal *Anas discors* and the Cinnamon Teal *A. cyanoptera* are kept together they interbreed freely, producing fertile hybrids, and forming in time a single mixed population.

The various types of isolating mechanism will now be discussed in turn.

II. A. *Isolation by Habitat*

Clausen described an example of isolation of this kind between two species of *Viola* in Denmark. *V. arvensis* has small flowers with yellowish-white petals and pinnate stipules, and is a common plant on chalky soils, whereas *V. tricolor* has large flowers with blue or violet petals and palmate stipules, and is found mainly on acid soils. Hybrids between the two forms are fertile, and in the second generation a wide variety of types are obtained with various combinations of the parental characters. But such intermediates are rare in nature and are confined to neutral or slightly acid soils. Thus the two species remain distinct because each can survive in soil conditions impossible for the other. This would involve an enormous wastage of gametes if crossing between the two species were at all common. But plants will usually be fertilized by pollen from close neighbours, growing on the same soil, and so of the same species, and hybridization will be common only in regions where the two kinds of soil border one another, and where the hybrids have a chance of surviving.

Such situations are not uncommon in plants. In Britain, the sea campion *Silene maritima* is a plant of pebble beeches, screes, and rock faces, and the related bladder campion *S. vulgaris* of fields. Although when crossed artificially the two species give fertile hybrids in the first and later generations, Marsden-Jones and Turrill found that few hybrids were formed in nature. A similar but vastly more complex picture has emerged from the study of the 250 or more 'species' of oak trees of the northern hemisphere. These species differ in size and habit of growth, in the shapes of their leaves and acorn cups, in being deciduous or evergreen, and also in the kinds of soil in which they thrive. Hybrids are common in nature, particularly on the borders between different types of habitat, and are often fully fertile. Clearly a considerable amount of genetic exchange between these species must occur, yet, despite the widespread occurrence of hybrids, it is possible to recognize, and so to name as distinct species, a number of characteristic forms of oaks.

The maintenance of this kind of isolation between plants depends partly on natural selection acting both against hybrid seeds and against seeds of one or other species which fall in an unsuitable habitat, and partly on the fact that crossing usually takes place between individuals growing close to one another. In animals individuals may travel much greater distances, so that matings between individuals originating in different habitats may be commoner. To this extent the maintenance of isolation by habitat alone may be less effective in animals than in plants. However, in animals such isolation may be reinforced in one of two ways. First, animals often select the habitats in which they live and reproduce, and second, they may select their mates.

In the three-spined stickleback, *Gasterosteus aculeatus*, Heuts described two distinct populations in Belgium, one living all the year round in fresh water, the other wintering in the sea and breeding in the summer in river estuaries. Although hybrids can be obtained by artificial insemination, they do not occur in nature, because the parental forms are confined to waters of different salinities during the breeding season. In this case the physiological tolerance of the two populations, and so the selection of breeding grounds, is determined genetically. Often, however, as was described in the last chapter, the selection of habitats by individuals is influenced also by their own past history. In either case, the selection by individuals of particular habitats will help to preserve the identity of different populations.

II. B. *Seasonal Isolation*

A clear example of isolation by a difference in the season of flowering was described by Whitaker in two species of *Lactuca*, both common weeds in the United States. One species flowers in early spring, the other in summer. Occasionally the flowering seasons overlap, and then hybrids are formed.

It seems to be rare, however, for seasonal isolation alone to separate two species.

II. C. *Isolation due to Mating Preferences*

In animals this is the most widespread process separating related species inhabiting the same geographical area. Even though two species may cross in the laboratory, this is often possible only because the individuals have been taken from their natural surroundings, and have been prevented for a considerable period from mating with a member of their own species.

Often the most striking differences between closely related species are in those characters whereby individuals can recognize sexual partners of their own species. The three species of leaf warblers in Britain can easily be recognized by the songs of the males in spring, but only with difficulty in other ways. In only one case are the songs of two British warblers difficult for a human being to distinguish, namely those of the Blackcap and Garden Warbler. It is a curious fact reported by Lord Grey that a Garden Warbler which nested in successive years in a quarry tolerated the presence in the same territory of Willow Warblers, but not of Blackcaps. It would be interesting to know whether females of these two species experience a similar difficulty in distinguishing the two songs. If so, interspecific matings would still be improbable, because in the Blackcap, alone among British warblers, the male has a plumage strikingly different from that of the female, and also from that of the male Garden Warbler.

These marked differences in sexual recognition characters suggest that natural selection may have been effective in emphasizing them. I shall return to this possibility later. But there is an interesting exception which helps to prove the rule, drawn from Lack's observations on the Galapagos finches. In most finches the male plumage is bright in colour and characteristic of the species, and is displayed during courtship. The plumage of the Galapagos finches is dull, and differs little from species to species. Nevertheless males will defend their territories against males of their own species, but usually tolerate males of other species. Lack was able to convince himself by observation, and by experiments with stuffed specimens, that the birds could recognize members of their own species primarily by the size and shape of their beaks, and that both territorial behaviour, in which two males grip beaks, and courtship, in which food is passed from the beak of the male to that of the female, serve to show off the characteristic beak shapes. It is natural that beak shape should serve as a recognition signal in these finches, since it is in this respect that the species differ most from one another. These differences have evolved, probably in the recent past, as adaptations to different diets. It seems that there has not yet been time for the evolution of differences in song and plumage, so that specific recognition must depend on a character which seldom serves this function in other passerine birds.

II. D. *Mechanical Isolation*

It is often the case in insects that the only, or the most noticeable, difference between two species is in the structure of the external genitalia. This led to the 'lock and key' theory of specific isolation, according to which the genitalia of a male will fit into those of a female of the same species as a key fits a lock, but will not fit the genitalia of females of other species. Observation has failed to confirm this theory. In *Drosophila*, for example, it may be rare for a male and female of different species to attempt copulation, because of difference in courtship behaviour, but if copulation is attempted the structural differences between the genitalia do not prevent effective fertilization. Although sometimes the crossing of animal species may be prevented by mechanical difficulties, such cases are unusual, and impediments to hybridization occur earlier, from behavioural differences, or later, from hybrid inviability or infertility.

Plants, however, lack a nervous system of their own, and differences in structure or colour of flowers are correspondingly important as isolating mechanisms, by exploiting the nervous system and sense organs of the insects which carry pollen. An extreme example will make this clear. Orchids of the genus *Ophrys* are pollinated by male bees of the genus *Andrena*. The flowers resemble female bees, and so attract male bees which attempt to copulate with them, picking up pollen in the process. Since different species of orchid resemble females of different species of bee, any one orchid species will tend to attract only one kind of bee, which will subsequently transfer the pollen he has collected to other orchids of the same species.

In *Ophrys* the prevention of interspecific pollination depends on genetically determined differences in response in different species of bees. The process is effective, but depends on an abundant and varied insect fauna. A less complete degree of isolation can result from the fact that individual bees form habits; that is to say, they visit in turn a series of flowers of the same shape, colour, or scent, having learnt that pollen or nectar is to be found there. Mather found that when two species of *Antirrhinum* were grown in alternate rows, very few hybrid seeds were formed, although the two species were cross-compatible. This was explained when he observed that an individual bee would usually confine its visits to one or other species.

In both the above examples, isolation depends on the recognition by insects of the form, colour, or scent of particular flowers. Isolation may also be due to structural differences between flowers which ensure that only particular kinds of insects are mechanically able to pick up pollen, or to transfer it to the stigma of another flower. In flowering plants which are pollinated by insects, these kinds of isolating mechanisms are of great importance, occupying a similar position to isolation due to mating preferences among animals.

II. E. *Prevention of Fertilization*

In many aquatic animals the ova and sperm are shed into the water, fertilization being external. In such animals, isolating mechanisms of any one of the kinds II A, B, or C above may be effective: spawning may take place in waters of different depths, salinities, or temperatures; breeding may occur at different seasons; spawning may be preceded by complex courtship displays, or, in sessile animals, the release of sex cells by all the individuals in a particular place may be triggered off by the discharge of a specific chemical substance by some of them. But even if such mechanisms break down, so that ova and sperm of different species are present together, interspecific fertilization may still be absent or rare, because of a lack of attraction between egg and sperm, or lack of penetrating power of the sperm.

In animals with internal fertilization, interspecific copulation and insemination is not always followed by the union of egg and sperm. Patterson has shown that in some interspecific matings in *Drosophila* the sperm fail to survive in the sperm receptacles of the females of other species. This seems to be an important method of isolation between *Drosophila virilis* and species closely related to it, but no such inviability of sperm has been observed in many other species crosses in the genus.

Similar physiological barriers to fertilization are common in flowering plants, in which the pollen tube must grow down the style of the female before fertilization can take place. Fertilization may fail because pollen from a short-styled species cannot grow a tube of sufficient length in a species with a long style, or because the growth of pollen tubes in foreign species is slowed down or prevented. Stebbins argues that such processes are seldom the primary cause of isolation between closely related species, because in most such cases, even if fertilization were to take place, the hybrids would be inviable or infertile. However, if pollen tube growth is inhibited, this is certainly the primary cause of isolation in the sense that it acts at an earlier stage in the reproductive process than would hybrid inviability; it is also different in its effects, in that it does not involve a wastage of seeds through the formation of inviable embryos.

III. *Hybrid Inviability*

It is often possible by artificial insemination, or, if fertilization is external, by artificial mixtures of ova and spermatozoa, to obtain hybrid zygotes between members of different families, orders, or even classes. Such hybrid embryos usually die at an early stage. Death or weakness of first generation hybrids is also not uncommon in crosses between closely related species, or even between geographical races of the same species.

Moore has made an extensive study of hybrids between twelve species of frogs of the genus *Rana* found in North America. In some cases no cleavage of the fertilized egg into a number of cells occurs; in many hybrids cleavage is normal, leading to the formation of a typical ball of cells, or blastula, but the process of 'gastrulation' whereby this is converted into a two-layered gastrula fails to take place; in still others gastrulation is normal, but development ceases at a later stage. Only in one group of frogs, including *Rana pipiens* and three other closely related species, are interspecific hybrids formed which develop normally into adults.

Rana pipiens has a particularly wide geographical distribution, covering a great range of latitudes from Canada to Central America. The species occupies a variety of different habitats, and shows considerable morphological variation throughout its range. More interesting, frogs from different latitudes show differences in physiology adapting them to high or to low temperatures. Embryos from southern populations can withstand higher temperatures during development than can those from Canada. But their rate of development in cold water (10 °C.) is slower than for the northern races. The rate of development of the embryos from all regions increases with temperature, but the increase is greater in embryos from the south, which at 30 °C. develop more rapidly than do northern embryos. Thus at low temperatures the rate of development is greater for the northern races, and at high temperatures for the southern ones. These differences in temperature tolerance and in developmental rate between races of the single species *Rana pipiens* exactly parallel differences between distinct species which in nature are confined either to high or to low latitudes.

Moore found that if he crossed individual *R. pipiens* from northern and southern populations the hybrids failed to develop, the degree of hybrid inviability being roughly proportional to the difference in the latitudes from which the parents were collected. Yet viable hybrids can be obtained between *R. pipiens* and members of other species from the same latitude. The relevance of these facts will be considered further on page 262, when discussing the genetics of hybrid inviability.

These observations show that, although there is a rough correlation between the extent of morphological divergence between populations and the degree of hybrid inviability, it is impossible to predict from a morphological comparison of two populations whether hybrids between them will be viable.

In species crosses it is often found that offspring of one sex only are obtained, embryos of the other sex failing to survive, or, if both sexes are obtained, that one or other sex is sterile. Haldane pointed out that when one sex is inviable or infertile, it is usually the heterogametic sex (i.e. the sex with two unlike sex chromosomes, X and Y, which therefore produces two kinds of gametes – see page 62). In most animal groups, including mammals and most insects, the male is the heterogametic sex, and it is common to find either that male hybrids are inviable, or that they are infertile while their female sibs are fertile. In the Lepidoptera (butterflies and moths) and in birds the female is the heterogametic sex, and in hybrids in both these groups there is often either an excess of males, or the females are sterile. Exceptions to 'Haldane's rule' are known, but it holds in the majority of cases; possible explanations of it will be discussed in the next chapter.

Just as it is possible by artificial insemination to obtain hybrids between species which normally will not mate, so it may be possible to obtain viable hybrids between species between which the hybrids normally die. If wheat flowers are pollinated with rye pollen, no embryos develop. However, Pissarev and Vinogradova dissected out wheat embryos from their endosperm, and grafted them into rye endosperm. The plants developing from such compound seeds resembled normal wheat in appearance, but differed in that, when pollinated with rye pollen, true hybrid embryos were successfully formed in up to 25 per cent of cases. This result has since been confirmed by Hall. Further research along these lines, in addition to its possible practical applications, may help to discover some of the reasons for hybrid inviability.

IV. *Hybrid Infertility*

The most famous of all animal hybrids, the mule, has a vigorous constitution but is sterile. Such sterility of one or both sexes in species crosses is a common phenomenon; it is also, for the student of evolution, a highly frustrating one, since genetic analysis of the differences between species requires a study of the second and later generations. Further examples of this phenomenon, and its possible causes, will be discussed in the next chapter. It is, however, clear that if the first generation hybrids between two species are sterile, this forms an effective barrier to the exchange of genetic material.

V. *Hybrid Breakdown*

Even though viable and fertile hybrids between two species are obtained, some barrier to exchange of genetic material may still exist if hybrids of the second generation, or from back-crosses of the F_1 hybrids to the parental species, are inviable or weakly. Such 'hybrid breakdown' is not uncommon. For example, Harland found that although the F_1 hybrids between the cottons *Gossypium hirsutum*, *G. barbadense*, and *G. tomentosum* are vigorous and fertile, a high proportion of the second generation seedlings failed to survive. In crossing *Drosophila pseudoobscura* and *D. persimilis* (see page 222) vigorous F_1 hybrids of both sexes are obtained. The male hybrids are wholly sterile, an example of hybrid sterility agreeing with Haldane's rule. It is therefore impossible to breed a second hybrid generation. However, an F_1 female can be back-crossed to either parental species. If this is done, it is found that the offspring are very weakly compared either to the parental species or to the F_1 hybrids.

This completes the review of isolating mechanisms between species; their evolution and genetics will be discussed in the next chapter. The classification given is based in the main on the stage during the process of reproduction at which the barrier is set. Three general points must be made in conclusion.

First, any two species are likely to be separated by more than one of these processes.

Second, there is an important distinction between mechanisms of the groups I and II, and those acting later. In the former cases, reproductive isolation is achieved without any wastage of gametes of either parental species; in the latter cases, F_1 hybrids are formed but do not leave descendants in future generations, so that the gametes of the parental species which unite to form such hybrids are wasted.

Finally, it does not follow that these isolating mechanisms have evolved in the same order in time as that in which they now act. As we shall see later, it is very probable that behavioural or mechanical isolating mechanisms are preceded, in evolution, by hybrid breakdown, sterility, or inviability.

CHAPTER 16

The Genetics of Species Differences

This chapter will attempt to answer two questions. First, how far is the genetic determination of the differences between species the same as that of the differences between members of the same species? If the Darwinian view, that varieties are incipient species, is correct, then we would expect to find that the genetic differences between species are similar in kind, though greater in extent, to the differences between individuals within a species. The second question concerns the origins of the isolating mechanisms described in the previous chapter. At first sight it may seem that these processes are different in kind from anything which can be found within species. So long as genetic studies are confined to a single species, mutant forms may be found, and in some cases they may be sterile or inviable, but if fertile they are normally not reproductively isolated, mating readily with other members of the species, and giving fertile offspring. We have therefore to ask whether there exist, for example between geographical races of a single species, inter-mediates between complete inter-fertility and complete re-productive isolation, intermediates which might give indi-cations of how the latter condition has evolved from the former. We have also to examine the genetic basis of such reproductive isolation, and to inquire how it has evolved, and what part has been played in its evolution by natural selection.

257

(i) *Mendelian Inheritance in Species Crosses, and Some Complications*

It was suggested in Chapter 3 that the great majority of genetically determined differences between the members of a species are caused by biochemical differences between chromosomes. When a particular difference in phenotype is due to a biochemical difference at a single chromosomal region, or locus we say, as a sort of shorthand, that it is due to a single gene. In practice, we reach this conclusion for any particular character, for example dumpy wings in *Drosophila* or albinism in mice, because the character 'mendelizes'; that is, because we get $3:1$ ratios in an F_2, $1:1$ ratios in the backcross of F_1 animals to the recessive parent, and so on.

The first question to be asked, then, is how far do the differences by which species are recognized behave in a similar way. As an example, an account will be given of the findings of Spurway on the genetics of inter-racial hybrids of the European crested newt, *Triturus cristatus*. Four geographical subspecies of this newt were recognized by taxonomists before any genetic study had been made. The subspecies found in Britain, *Triturus cristatus cristatus*, extends across most of northern Europe. These newts have a rough toad-like skin. The belly is orange or yellow, with black spots which tend to aggregate to form dark bands laterally. Just dorsal to these dark bands there are bands of white stipples. The throat is black, with a number of small white spots.

The Italian subspecies, *T.c. carnifex*, differs in having a smooth skin, a pattern of belly spots which show no tendency to aggregate, relatively few white stipples on the flanks, and a black throat with only minute white specks. A third subspecies, *T.c. danubialis*, from the Danube valley below Vienna, differs from the two previous forms in its small size, slender form, and well-developed neck. This subspecies is polymorphic for the differences in colour and texture which distinguish *cristatus* and *carnifex*. Finally, a fourth subspecies, *T.c. karelinii*, is found in south Russia and Asia Minor. These newts are larger than the other races; the skin is rough, though less so than in *cristatus*; white stipples are present on the flanks; the black belly spots form an irregular marbling over the belly, but in at least some individuals they first appear in longitudinal bands along the flanks. The most characteristic features of the coloration of *karelinii*, however, are the presence of a blue sheen, and the fact that the yellow-and-black belly pattern extends forwards over the throat, there being no distinct black or grey throat as in the other subspecies.

In Spain there is another species, *Triturus marmoratus*, of a general green-and-black coloration. This species extends into France, where its range overlaps that of *T.c. cristatus*.

These differences in form, skin texture, and coloration between the four races of *T. cristatus* have been considered sufficient to justify classification into separate subspecies. As will be described below, there is considerable hybrid infertility and hybrid breakdown in crosses between the subspecies, similar in kind though less in extent to that observed in the species cross *T. cristatus* × *T. marmoratus*. Nevertheless, it has been found that many of the differences which have been used in the recognition of these sub-species depend on single gene differences, as evidenced by the Mendelian segregations obtained.

For example, the 52 F_1 animals between *cristatus* and *carnifex* had throats with the distinct white spots characteristic of *cristatus*. From a backcross of these F_1 animals to *carnifex*, 18 offspring were obtained with the *cristatus-type* throat and 15 with the black throat characteristic of *carnifex* – a satisfactory fit to the Mendelian 1:1 ratio. These results are explained if *cristatus* has the genotype T^{cr}/T^{cr}, and *carnifex* is homozygous for a recessive allele, t/t. The F_1 would then be T^{cr}/t, and resemble *cristatus*, and the backcross generation would contain 1 T^{cr}/t (resembling *cristatus*) to 1 t/t (resembling *carnifex*).

Using this kind of argument, Spurway was able to suggest the following genotypes for the various subspecies, accounting for most of the taxonomic differences between them in terms of only five loci, as shown in the table on page 260.

Two points about these results should be emphasized. First, the expression of the various genes is not always identical in the different subspecies; for example, the presence of the dominant allele, *L*, for the lateral aggregation of the belly spots is always

	PHENOTYPE					GENOTYPE
Subspecies	Throat	White stipples	Belly spots	Blue sheen	Skin	
T.c. cristatus	Black with white spots	Many	Aggregated laterally	Absent	Rough	$T^{er}T^{er}\ SS\ LL\ bb\ R^{er}\ R^{er}$
T.c. carnifex	Black with minute white specks	Few	Not aggregated	Absent	Smooth	$tt\ ss\ ll\ bb\ rr$
T.c. danubialis	Black with spots or specks	Variable	Variable	Absent	Rough or smooth	$t\quad s\quad l\qquad r$ or or or bb or $T^{er}\ S\ L\quad R^{er}$
T.c. karelinii	Belly pattern continues over throat	Many	Aggregated laterally at first appearance in some individuals	Present	Somewhat rough	$T^k T^k SS$ or $BB\ R^k R^k$ L

recognizable in adult *cristatus*, but is masked in adult *karelinii* by a general extension of the black pigment over the belly, and can be recognized only in newly metamorphosed individuals. The second point, which follows from this, is that differences at these five loci, although responsible for most of the colour differences by which the subspecies were recognized, constitute only a small part of the genetic differences between the races. This will become clearer when the fertility of the hybrids is discussed.

A rather similar picture emerged from Danforth's investigation of interspecific and intergeneric pheasant hybrids, and from a study by Clarke and Sheppard of hybrids between several species of swallowtail butterflies. In all these cases it seems that at least some of the most striking differences between species are due to genes at a few loci, but that these loci account for only a small part of the genetic differences. Species probably differ in the alleles present at very many loci – perhaps thousands – but only in a few cases does the replacement of one allele by another produce an effect so large that it can be recognized individually, and Mendelian ratios obtained.

It is often found that the F_1 hybrids between two species are intermediate in most respects between the parents, and that the F_2 is highly variable, consisting of a wide range of types with little sign of Mendelian segregation. There is nothing surprising about this. It is the result to be expected when the characters by which the parental forms differ are determined, not by a single gene difference, but by a large number of such differences, each of small effect.

Difficulties arise when it is found that the hybrids between species are inviable or infertile, either in the first or later generations. It has sometimes been argued that hybrid inviability and infertility could not arise by the gradual accumulation of gene differences of the kind which exist within populations, and hence that some special processes must be involved in the origin of a new species. One proponent of this view was Goldschmidt, whose arguments are discussed further on pages 316–20. More recently Carson has put forward a somewhat similar view, based on his studies of the *Drosophila* of Hawaii. I find their arguments muddled and unconvincing.

The first point to make is that inviability and infertility are not found only in hybrids between species; they also occur in the hybrids between geographical races of the same species. Hybrids between races of *Rana pipiens* are inviable, as described on page 254. In the case of the races of the newt *Triturus cristatus* described in this chapter, the F_1 animals are very vigorous but have a greatly lowered fertility. Such cases are by no means unusual in hybrids between geographical races. Of course it is always open to one to insist that when two populations show any degree of hybrid inviability or infertility they are different species, and then to claim that species hybrids show a special peculiarity not found within species; but clearly such an argument would be illegitimate. The essential observational point is that all degrees of compatibility exist between populations, from fully viable and fertile hybrids to hybrids which die early in development. The theoretical question then arises, can we account for this without invoking some special process? I think we can.

Suppose we start with a diploid population whose genotype at two loci is AB/AB; it turns out that we do not have to consider more than two loci, although of course more than two would be involved in any actual case of speciation. This population is divided into two geographically isolated parts. One of these remains unchanged. The other, in a changed environment, evolves first into $A'B/A'B$, and then to $A'B'/A'B'$. Clearly this requires that in the new environment it was advantageous to replace A by A', and having done so to replace B by B'. Suppose, however, that it is only advantageous to replace B by B' if A has already been replaced by A', but that in an AA genotype B' would be disadvantageous. It is easy to imagine cases in which this would be so.

Now consider what happens among hybrids. F_1 hybrids will be $AB/A'B'$, and in later generations genotypes such as AB'/AB' will arise. Whether these hybrids show inviability in the F_1 or breakdown in later generations will depend on the dominance relations; if A is dominant to A' but B' to B, the F_1 will be inviable, but in most other situations inviability will only appear in later generations.

The point of this simple-minded example is to show that it is quite possible for two populations to diverge under the influence of natural selection, and for their hybrids to show inviability. Dobzhansky has called the relationship between sets of genes which work together 'coadaptation'; in our example, AB, $A'B'$ and $A'B$ are coadapted, but in AB' coadaptation breaks down. Clearly the larger the number of genetic differences between two populations, the more likely it is that coadaptation will break down. The example also shows clearly why it is that there may be breakdown in later generations even when the F_1 hybrids are viable.

The concept of coadaptation also explains 'Haldane's Rule', according to which it is usually the heterogametic sex which is either inviable or infertile in the F_1 between two species. Using capital and small letters, not for individual genes, but for the chromosomes present in the two parental populations, and considering an example in which the male is the heterogametic sex, we can describe the males of the parental species as $ABCX/ABCY$ and $abcx/abcy$ respectively, where A, a, B, b,C, c, stand for autosomes, and X,x, Y,y, for sex chromosomes. F_1 hybrid females have the genotype $ABCX/abcx$; they have two complete 'teams' of chromosomes, one proper to each parental species. F_1 males have the genotype $ABCX/abcy$, or $ABCY/abcx$. Now it is usually the case that the Y chromosome is genetically inactive. These male hybrids can therefore be written $ABCX/abc$, or $ABC/abcx$; it is now apparent that they carry one complete 'team' of chromosomes from one parent, but an incomplete team from the other. Thus the inviability or infertility of the heterogametic sex (in this case the males) in first generation hybrids can be accounted for in the same way as can the inviability of second generation hybrids.

It follows that there is no need to propose any special process to account for hybrid inviability. Two reservations are needed. First, there may sometimes be active selection in favour of hybrid inviability (see the discussion of cotton hybrids on page 273); second, special difficulties arise in explaining some types of structural differences between the chromosomes of different populations (see the discussion of newt hybrids on page 275).

(ii) *The Genetics of Hybrid Infertility*

Meiosis, the process of cell division whereby haploid gametes are produced from diploid cells, differs in one important respect from other developmental processes. It depends for its proper functioning on a certain degree of similarity between the two sets of chromosomes, of maternal and paternal origin, present in the diploid individual. If homologous chromosomes do not pair in meiosis, gametes with irregular chromosome numbers are produced; such gametes are called 'aneuploid'.

In species or race hybrids, the dissimilarity between the chromosomes of paternal and maternal origin may cause irregularities in one of two ways, illustrated in Figure 24. The chromosomes may fail to pair; if so, single unpaired chromosomes will pass to one or other daughter cell more or less at random. Consequently, aneuploid gametes, with some kinds of chromosome present twice and other kinds wholly absent, are produced. Zygotes to which aneuploid gametes have contributed usually die at an early stage, so that individuals producing only such gametes are effectively sterile.

A second kind of irregularity arises in hybrids in which homologous chromosome regions pair, but in which the parental species differ by one or more translocations (see page 128). As shown in Figure 24, meiosis in hybrids carrying translocations produces gametes which may contain some chromosome regions in duplicate, and others not at all. In other words, the result is again the production of aneuploid gametes.

Both these kinds of irregularities have been shown by Callan to occur in hybrids between different races and species of crested newts, whose genetics was described above. In all cases the hybrid males are less fertile than the females. In spermatogenesis of hybrids between different races of *Triturus cristatus*, the number of chiasmata formed is lower than in the parental races, and in most cells some unpaired chromosomes can be seen. The male hybrids between the species *T. cristatus* and *T. marmoratus* are wholly sterile; in spermatogenesis the majority of chromosomes are unpaired. There is also evidence that the species and races differ by translocations, since occasional quadrivalents

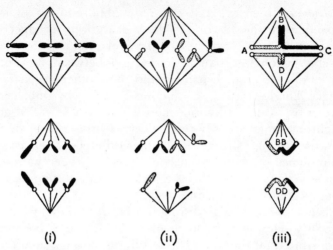

(i)   (ii)   (iii)

Figure 24. Diagrams of the first meiotic division; (i) in a normal diploid with three pairs of homologous chromosomes, which in the early stages of division lie side by side, attached to the 'spindle' at a particular point, the centromere, indicated in the diagrams by a white circle. Later the two centromeres of a pair move to opposite poles of the spindle, so that at each pole a complete haploid set of chromosomes is collected. (ii) in a hybrid in which three paternal chromosomes (hatched) do not pair with the three maternal chromosomes. (iii) in a hybrid in which the parental chromosome sets differ by a translocation.

(Each chromosome in these diagrams consists of two threads. Thus a pair of chromosomes, or bivalent, as shown in (i), has the structure shown in Figure 4, top right. The two threads composing a chromosome are separated in the second meiotic division. These complications have been omitted from the present figure, since they are not necessary to an understanding of chromosomal infertility.)

(four chromosomes pairing during meiosis, as in Figures 16, 24(iii)) can be seen. It should be pointed out that the absence of a visible quadrivalent in a cell undergoing meiosis does not prove that no translocations are present, since the relevant chromosomes may fail to pair.

It seems, therefore, that the sterility of male species hybrids, and the lowered fertility of inter-racial hybrids, arises in part because chromosomes fail to pair in meiosis, and in part because the parental forms differ by translocations, the consequence of both these facts being the production of aneuploid gametes.

However, these are not the only abnormalities of spermatogenesis in these hybrids. Many of the cells produced by meiosis degenerate and do not develop into mature sperm. There is no obvious causal connection between the abnormalities of chromosome pairing and these other degenerative changes. This leads us to a distinction suggested by Dobzhansky between two kinds of hybrid infertility, which he called genic and chromosomal sterility respectively. Chromosomal sterility may arise because structural or other differences between the chromosomes of the two parents prevent proper pairing in meiosis; genic sterility reflects a lack of coadaptation between the genes controlling meiosis.

It is not always easy to distinguish between these two causes of sterility. However, in some plant hybrids it is possible to demonstrate that the cause of infertility is chromosomal rather than genic; an example is afforded by the analysis by Karpechenko of hybrids between radish and cabbage.

Both radish and cabbage have a diploid number of 18, so that F_1 hybrids also have 18 chromosomes, 9 from each parent species. In the meiosis of F_1 hybrids no chromosome pairing takes place, so that the vast majority of gametes are aneuploid. The hybrids are therefore almost completely sterile. However, an occasional gamete is formed containing a complete (diploid) set of 18 chromosomes resembling those of the somatic cells of the hybrid. The union of two such exceptional gametes gives rise to viable F_2 plants with 36 chromosomes, two complete sets from both radish and cabbage. Such plants are called tetraploids – when, as in this case, the chromosomes are derived from different species, allotetraploids. These hybrids are fully fertile. In meiosis, 18 normal bivalents are formed, 9 composed of pairs of homologous radish chromosomes and 9 of cabbage chromosomes. Fertility is restored in the allotetraploid because each chromosome now has a partner with which it can pair. The evolutionary importance of allotetraploids will be discussed in the next section.

The restoration of fertility in allotetraploids depends on the fact that each chromosome has one and only one partner with which it pairs in meiosis. This explains a rule which Darlington showed to hold for many plant hybrids. In those diploid F_1 hybrids in which little pairing takes place, and which are therefore infertile, the corresponding allotetraploid is usually fertile; if there is good pairing and fertility in the diploid hybrid, the allotetraploid is often infertile. In the latter case, pairing in the tetraploid is often irregular, since each chromosome has three possible partners.

Allotetraploidy provides a mechanism whereby a new species can arise suddenly, without a long period of geographical isolation.

(iii) *Species Formation by Hybridization and Polyploidy*

Hybridization may give rise at a single step to a new species, distinct from either parental form. In the evolution of the genus *Crepis* in America, many asexually reproducing species hybrids have established themselves in nature. But these forms are not dignified with specific names, and their method of reproduction makes it unlikely that they will have a long evolutionary future. The origin of a new sexual species by hybridization requires that the hybrids should be fertile when mated to each other, but that if backcrossed to either parental species the offspring should be inviable or infertile. Such hybrids can arise by a doubling of the chromosome number (page 266).

This process has been very common in the evolution of plants, but rare or absent in animals. The reasons for its rarity in animals are not fully understood, but two suggestions can be made. First, many plants are hermaphrodite and so possess no genetic mechanism determining that some individuals shall be males and some females, whereas in most animals the sexes are separate and a genetic sex-determining mechanism exists. Now a doubling of the chromosome number may upset an X-Y sex-determining mechanism (see page 62); the reasons for this will not be given here, but readers who find genetic arguments easy will be able to work them out for themselves. A second cause for the rarity of allopolyploid species of animals may lie in the different patterns of development in plants and animals. The growth of plants is indeterminate, giving rise to individuals with varying numbers of branches, leaves, roots, etc. A common consequence of polyploidy is an increase in size and a change in the relative sizes of different parts. Such changes are more likely to produce a viable and efficient organism if development is indeterminate in the above sense, than if, as in most animals, it leads to the appearance of a fixed number of parts with fixed relations to each other.

Whether or not these reasons are adequate to explain the rarity of allopolyploid species of animals, such species are certainly much commoner among plants. Two examples will be described, in two important agricultural plants, wheat and cotton.

Wheat species occur with somatic chromosome numbers of 14, 28, and 42; they are diploid, tetraploid, and hexaploid respectively, containing two, four, and six chromosome sets. The tetraploid emmer or durum wheats have evolved from the wild tetraploid species, *Triticum dicoccoides* of Syria. The origin of this species is uncertain, but since it has twice the chromosome number of some other wheat species, for example *T. monococcum*, it is possible that it arose as an allotetraploid hybrid between two diploid species. More remarkable is the origin of the 42-chromosome bread wheats, *Triticum vulgare*. It was long suspected that these wheats were allopolyploid hybrids between the tetraploid *T. dicoccoides* and some other species with 14 chromosomes. The discovery of a technique of producing polyploids experimentally, by treatment with colchicine, has made it possible to confirm this suspicion, and to identify the other parent.

This parent belongs to the genus *Aegilops*, a plant of no economic value which grows as a weed on the borders of wheat fields in the Near East. *Aegilops* has a somatic chromosome number of 14. F_1 hybrids between *T. dicoccoides* and various species of *Aegilops* have 21 chromosomes in somatic cells, 7 from *Aegilops* and 14 from *Triticum*. Artificially produced allopolyploid hybrids had a somatic chromosome number of 42, were fully fertile when crossed together, resembled bread wheats in appearance, and when crossed to these wheats gave fertile hybrids. Thus it has been possible to repeat experimentally the process whereby the bread wheats evolved. The original hybridization which gave rise to them probably occurred between cultivated emmer wheat and a species of *Aegilops* growing as a weed in cornfields.

This example illustrates the use of a technique for investigating the past evolutionary history of a species. If, in a hybrid, the chromosomes from the two parents pair, i.e. lie side by side, in meiosis, that is taken as evidence that the chromosomes of the two parents are derived from some common ancestor in the fairly recent past. Thus in the offspring of a cross between bread wheat and artificial *Aegilops-T. dicoccoides* hybrids, the 7 *Aegilops* chromosomes from the latter parent paired in meiosis with 7 chromosomes from the bread wheat parent; this is taken as proving that 7 of the chromosomes in bread wheats have been derived in the recent past from *Aegilops*.

Another application of this technique will now be described. In both the New and Old Worlds there are diploid species of cotton with 13 pairs of chromosomes in the somatic cells. In the New World there are also tetraploid species with 26 pairs of chromosomes. These tetraploids existed before cotton was domesticated, and it is fairly certain that they originated by hybridization and subsequent polyploidy between two diploid species, one of New and one of Old World origin. The evidence is as follows: artificial allopolyploid hybrids between New and Old World diploid cottons have been produced experimentally, and proved to be sufficiently fertile to give hybrid offspring when crossed to New World tetraploids. In these offspring, meiosis was fairly normal, most of the chromosomes pairing satisfactorily. As in the case of bread wheat, the process which gave rise to the tetraploid species has been repeated experimentally, although the precise species chosen probably differed from the original parents.

The importance of hybridization in the evolution of plants arises partly because the conditions which promote hybridization also create new habitats in which the new varieties or species produced can establish themselves. Anderson and Stebbins have recently discussed some of the great revolutions in the world's flora from this point of view. Of all such revolutions, the most dramatic was the replacement of the gymnosperms (conifers and their relatives) by flowering plants during the middle of the Cretaceous period. These writers suggest four events which may have contributed to this revolution: the retreat of the seas, leaving the coastal plains open to colonization; the evolution of the large herbivorous dinosaurs, and the consequent over-grazing of the gymnosperm flora; the evolution of the birds, which transport seeds over long distances; and finally, the evolution of the bees and of other insects which pollinate flowers. The first two of these factors would create new habitats for the evolving flowering plants, and all would promote hybridization, either directly, or because they would cause previously isolated populations to grow side by side. Now a study of the chromosome numbers of flowering plants makes it certain that allopolyploidy played an important part in their diversification, and the same is probably true of introgressive hybridization, although it is more difficult to obtain direct evidence on this point.

This picture is an attractive one, although it is open to one criticism. The appearance of the great herbivorous dinosaurs preceded by some millions of years the emergence of the flowering plants as the dominant flora. Still more puzzling is the fact that the change in the world's flora was not immediately followed by any major change in fauna, at least as far as land vertebrates are concerned. The herbivorous dinosaurs survived the flora change, and continued as the dominant land animals in a world clothed in flowering plants until the end of the Cretaceous period. They are not found in rocks of slightly later age; no really satisfactory explanation has yet been given for their final and rather sudden disappearance. It was only after a further period of millions of years that they were replaced by a fauna of large herbivorous mammals.

(iv) *Natural Selection and the Origin of Isolating Mechanism*

The inviability of hybrids between frogs of the species *Rana pipiens* from different latitudes, adapted to different temperatures, shows that an isolating mechanism can arise as a by-product of genetic adaptation of the parental populations to different conditions. The same is doubtless true of hybrid infertility and hybrid breakdown. What will happen if two populations, which have evolved isolating mechanisms of one of these kinds while geographically isolated from each other, subsequently spread into the same area? Since little or no exchange of genetic material can take place, merging of the two populations is unlikely. There are, therefore, two possible results; one or other population may become extinct in the area of overlap, or the two populations may continue to live side by side as distinct species.

In the latter case, animals which mate with, or plants which are fertilized by, members of the other species will contribute nothing to future generations; they will be less fit in a Darwinian sense than those which mate with members of their own species and leave fertile offspring. Consequently there will be a strong selection pressure tending to transform isolating mechanisms acting through the infertility or inviability of the hybrids into mechanisms which prevent the formation of such hybrids; that is, there will be strong selection favouring isolation by habitat, breeding season, mating preferences, or floral differences.

This argument, which is due to Dobzhansky, is a very convincing one, although at present there is little direct evidence that it is correct. What evidence there is comes from the study of *Drosophila pseudoobscura* and its relatives. In crosses of this species with the related *D. miranda* the F_1 hybrid males have a low viability and the females, although viable, are sterile. The two species overlap in nature over part of their range. Dobzhansky and Koller found that the degree of isolation due to mating preferences was greater for *D. pseudoobscura* taken from regions where the two species occur together than for those caught in regions distant from those inhabited by *miranda*. This suggests that natural selection has strengthened the behavioural isolating mechanisms in the areas of overlap, which are the only areas where such selection could act.

More direct evidence has been provided by an elegant experiment by Koopman on *D. pseudoobscura* and *D. persimilis*. No exchange of genetic material between these species can occur, because of the sterility of the F_1 males and the inviability of the offspring of F_1 females. In nature, *persimilis* has a restricted range entirely contained within that of *pseudoobscura*. Large-scale studies of the salivary gland chromosomes of these two species have failed to detect any hybrids in nature, although such hybrids are viable and could be recognized by this technique if they occurred. Therefore in nature the two species are separated by mechanisms acting before fertilization. These depend in part on the selection of slightly different habitats by the two species, and in part on the fact that recently captured strains show some degree of sexual isolation.

This sexual isolation, however, breaks down at low temperatures; in a mixed population kept in the laboratory at 16 °C. a high proportion of hybrids were found. Koopman decided to see if he could strengthen sexual isolation at 16 °C. by artificial selection in a population cage. He used strains of the two species homozygous for two different recessive mutants, so that he could easily recognize the hybrids, which alone were wild-type (i.e. did not show the effects of either mutant). He followed a number of generations of a mixed population of the two species, removing in each generation all hybrids formed (this latter measure was not strictly necessary, since in any case the hybrids could leave no progeny, but it simplified the analysis of his results). The proportion of hybrids formed in any generation provided a measure of the proportion of inter-specific as opposed to intraspecific matings which had occurred among the parents; the fewer the hybrids, the greater the degree of sexual isolation between the two species. In three parallel experiments, the first generation contained respectively 36, 22, and 49 per cent of hybrids. In the fifth and later generations the proportion of hybrids fell to 5 per cent.

This experiment demonstrates that artificial selection can increase the behavioural isolation between two species very rapidly. But it must be remembered that these particular species are in nature isolated by other mechanisms as well.

There is a need for further studies on the influence of mating preferences on the isolation of species in nature. But one general observation is suggestive. In animals in which the male can mate many times but the female once only, at least in a given season, it is often found that males are promiscuous and females selective in behaviour. For example, in *Drosophila subobscura* Milani found that males will attempt copulation with blobs of wax of the right size which have been moved in an appropriate manner, whereas the females, as described in Chapter 12, are highly selective. This is understandable, since females, but not males, would be effectively sterilized by an interspecific mating. However, as I have argued above, I believe that in this case selection has acted to ensure that females mate with the most fertile members of their own species, and not primarily to prevent interspecific matings.

The inviability of F_1 hybrids can arise as a corollary of genetic divergence, without selection in favour of inviability as such. The following observations of Stephens on cotton hybrids, however, suggest that there are circumstances in which natural selection may be directly responsible for hybrid inviability. Hybrids between the cotton species *Gossypium hirsutum and G. barbadense* are viable and fertile, but a large proportion of the F_2 or back-cross progeny are inviable (see page 255). In some cases, however, the F_1 hybrids are weak, with the stems, petioles, and leaf midribs covered with a layer of cork. This syndrome, known as 'corky', occurs if the hybrid carries a pair of complementary alleles, one from each parent species. The important observation is that the alleles responsible for the syndrome are found only in the zone of overlap of the two species in the West Indies and the north of South America. This suggests that selection in favour of genes causing hybrid inviability is acting in this area.

Such genes will be favoured by selection only if they increase the contribution to future generations made by plants carrying them. This would be true for genes which prevented interspecific fertilization, and so reduced the wastage of gametes for which such fertilization is responsible, but why should it be true of genes which act only after such wastage has occurred? A possible explanation is as follows: if the F_1 hybrids are vigorous, and if in general, they establish themselves close to the parental plants, then any fertile offspring produced by intraspecific fertilization of the same parental plants will grow in competition with these hybrids, and so will have less chance of surviving than they would have had were the F_1 hybrids inviable. If this argument is correct, then selection for hybrid inviability will be most effective in plants and sessile animals with relatively poor dispersive powers, since in such groups the offspring of a given individual compete with one another.

(v) *The Adam and Eve Effect*

One last difficulty remains to be discussed. I argued on page 262 – that no special processes are required to account for the origin of hybrid inviability or infertility. There is, however, a problem if it turns out that one population is homozygous for a gene (or chromosome structure) A and the other for A', and that the heterozygote AA' is necessarily less fit than either homozygote. In such a case it is difficult to see how the allele A' could increase in an AA population or the allele A in an $A'A'$ population. Such a situation is probably rare for alleles at a single locus, although it may hold for the genes determining the *Rhesus* antigens in man. It is, however, common for structural changes in chromosomes. For example it was shown in Figure 24 that an individual heterozygous for a chromosome translocation is likely to produce aneuploid gametes. We have seen that species and subspecies of the newt *Triturus* differ by translocations. Such differences are commonly found between species. How then can a translocation become established if the first individuals in a population with the translocation are infertile?

A possible solution of this difficulty has been suggested by Sewall Wright. He supposes a situation in which there are many localities in which colonies of the species are intermittently dying out, to be started again from single gravid females, or at most from a very small number of immigrants. If one of the few founders of such a colony carried a translocation, then the 'new' translocated chromosomes would, in that one colony, no longer be greatly outnumbered by the 'old' chromosomes. There would be a fair chance of fully fertile individuals homozygous for the new chromosomes being born. Such a population would be highly unstable, and either the new or the old types of chromosome would soon be eliminated from it. Should the new chromosomes be established in such a colony as the 'normal', then any further immigrants into the area, carrying the old chromosomes, would be at a severe selective disadvantage. To put the matter another way, whichever kind of chromosome is rare in the population will be the kind to be eliminated. A change-over from one kind to the other has a chance to happen only when a new colony is founded by a translocation heterozygote, so that the new and old kinds of chromosomes, both rare, start on a more equal footing.

Spurway believes that such a process may have occurred in the recent evolution of *Triturus*. She points out that newts live in colonies (in ponds) of the kind postulated by Wright, not only at the edges of their range but also at the centre of it. However, the process suggested by Wright is more likely to have occurred at a period when the population was extending its range. For example, the colonization of Italy from the north would involve crossing innumerable barriers from valley to valley and from pond to pond. At each of these steps there would be a small but finite chance of a new translocation being established. In such a case, reproductive isolation between two populations can arise at a single step, as the first stage in the evolutionary divergence of the populations; this may in turn initiate later morphological and physiological differentiation. Spurway has suggested the term 'Adam and Eve' speciation for this process, depending as it does on the peculiarities of the founder members of new populations.

CHAPTER 17

The Fossil Evidence

It is the job of this book to explain the past in terms of processes known to be going on in the present. So far we have considered what can be learnt from a study of the variation of animals and plants, their present geographical distribution, and a genetic analysis of the differences between them. These methods of study are effective mainly in elucidating relatively small-scale evolutionary changes; for example, genetic analysis can be applied only if two organisms are sufficiently closely related to cross and to produce viable offspring. We must now turn to the more direct evidence concerning the past provided by fossils, and inquire whether the patterns of change revealed are of a kind which can be explained by the known laws of variation, selection, and inheritance.

(i) *Rates of Evolution*

The simplest question which can be asked concerns the rates at which evolutionary changes have proceeded. There are various possible ways of measuring such rates, of which three examples will be given. The most direct is to determine the rate of change with time of some measurable characteristic in a known phylogenetic line. An instructive example is provided by changes in the shape of the grinding teeth during the evolution of the horses, of which abundant fossil material exists. Modern horses differ from their ancestors in the Eocene in that their grinding teeth are much taller relative to their width. This change has an adaptive significance in a grass-eating mammal. Grass is a siliceous material, and chewing it rapidly wears away

276

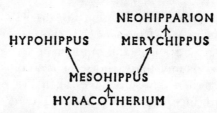

Figure 25. Relationships of five horse genera.

the surface of the teeth. Consequently the greater height of the teeth of modern horses is necessary if the teeth are not rapidly to wear down to the gum.

Simpson has measured the height and length (parallel to the tooth row) of unworn molar teeth in five genera of fossil horses, and calculated the mean value of the ratio of height to length for each. The relationships of the five genera are shown in Figure 25.

It will be seen that *Mesohippus* gave rise to two different lines of descent, one leading to *Hypohippus* and the other to *Merychippus* and *Neohipparion*; there are many other 'branches' in this phylogeny which do not here concern us. Knowing the ratio of tooth height to length in each genus, and the approximate period in millions of years which elapsed during the evolution of one form from another, it is possible to calculate the rates of evolution as percentage changes in this ratio per million years. The results are as follows:

	Percentage Change in the Ratio of Tooth Height to Length, per Million Years
Hyracotherium–Mesohippus	0·9
Mesohippus–Hypohippus	1·1
Mesohippus–Merychippus	4·9
Merychippus–Neohipparion	5·5

There are several points of interest here. First, even the most rapid change found, 5·5 per cent per million years, is very slow compared to the rates of change in single characters which can be produced by artificial selection in the laboratory. For example, it was shown in *Drosophila* that changes of the order of 3 per cent per generation in the number of bristles can be so produced. If we assume that a horse generation takes on the average 5 years, then it would have taken just over 100,000 generations to produce a 3 per cent change. Thus even the rather rapid evolutionary change during the transition from *Merychippus* to *Neohipparion* was some 100,000 times slower than changes which can be observed in experimental conditions.

Another point shown by the table is that although the direction of evolutionary change was the same throughout, its rate was not. In the transition *Hyracotherium – Mesohippus – Hypohippus* the rate of change was low, but there was a considerable acceleration of rate in the line leading to *Neohipparion*. A similar accelerated rate occurred in other lines of descent from *Merychippus*, including that leading to present-day horses, zebras, and asses of the genus *Equus*. Why this sudden acceleration in the rate of evolution? There is little doubt that it arose because *Merychippus* and its descendants abandoned the habit common to all earlier horses of browsing on leaves, and took the newly evolved grasses as their main food. Other horse lineages, which although now extinct, survived alongside the grazing horses for many millions of years, continued to browse on leaves, and in these there was no increase in the rate of evolution of tooth shape.

Evolution often involves qualitative changes whose extent is difficult to measure, and whose rate cannot therefore be so simply calculated. One method of comparing the rates of such changes was used by Westoll in a study of lungfish. The three surviving genera of lungfish are all adapted to life in stagnant water, and two of them can survive by burying themselves in the mud should the pool which they inhabit dry up completely. Their ancestry can be traced back for over 300 million years. During this period there have been changes in the skull, teeth, fins, and scales. Westoll listed 21 characters in which modern lungfish differ from their earliest known ancestors. By recognizing various stages of change in each of these characters, it is possible to calculate a total 'score' for each fossil genus, such

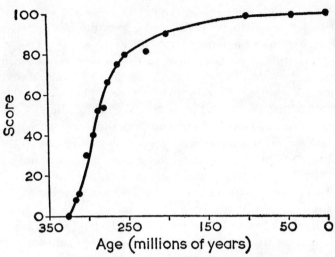

Figure 26. The rate of evolution in lungfishes (after Simpson, from Westoll's data). For explanation see text.

that the score would be 100 for a genus modern in every respect, and 0 for a genus primitive in every respect. In Figure 26 the scores for all known genera are plotted against their ages in millions of years. A smooth curve can reasonably be drawn through these points; the steep slope of this curve between 250 and 300 million years ago shows that lungfish were evolving relatively rapidly during that time, whereas evolution during the last 200 million years has been extremely slow.

This pattern, consisting of relatively rapid change soon after the origin of a major group, followed by a long period during which little or no change takes place, is a common one, though by no means universal. For lungfish, it is probably to be explained along the following lines. The early period of rapid change was one during which adaptation to their peculiar mode of life in a restricted habitat of stagnant water, liable to dry up, was being perfected. (At the same time still more rapid evolutionary changes were occurring in the descendants of the related osteolepid fishes, whose response to the drying up of a pool was, not to bury themselves in the mud like lungfish, but to travel overland to another pool. It was from this group that the amphibians and all later land vertebrates evolved.) Once the lungfish had evolved these adaptations, selection has acted to preserve rather than to change them.

In this example a measurement of the rate of evolution is based on a judgement, to some extent subjective, of the relative importance of a series of qualitative changes. The remarkable smoothness of the curve in Figure 26 is a justification of the approach A similar method depends on the rates at which genera arise or die out, a method which is also subjective to the extent that there are no hard and fast rules whereby to decide whether two fossils or sets of fossils should be placed in different genera. In a rapidly evolving group the length of time for which a particular genus will survive is likely to be relatively short, since it is likely either to become extinct in face of competition from a more recently evolved genus, or to be itself transformed in the course of evolution into a new genus. In fact the duration of survival of individual genera in a group will tend to be inversely proportional to the rate of evolution of that group.

The survival times of individual genera can be calculated either from extinct or from surviving genera. The survivorship of an extinct genus is the time in, say, millions of years during which representatives of that genus were fossilized, and of a surviving genus the time in the past at which that genus first appears in the fossil record. One way of presenting the results of such calculations is shown in Figure 27, for bivalved molluscs and for carnivorous mammals. These curves are to be inter-preted as follows: for extinct genera of bivalves, for example, about 50 per cent survived for 50 million years or more, about 20 per cent for 100 million years or more, and none for more than 275 million years, whereas for existing genera of carnivores less than 20 per cent were present 10 million years ago, and so on.

Two interesting deductions can be made. First, genera of carnivorous mammals, both surviving and extinct, have lasted on the average for a much shorter period than have genera of bivalves. This would be expressed more accurately by saying that the amount of difference between two carnivores which we regard as sufficient to justify placing them in different genera

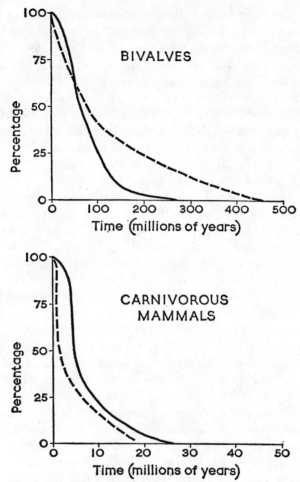

Figure 27. Survivorship curves for genera of bivalves and of carnivores (after Simpson). The continuous lines show the percentage of extinct genera which survived for various periods, and the broken lines the percentage of living genera which have already survived for various periods.

has taken less time to evolve than has the equivalent difference in bivalves; in this sense it is true to say that carnivores have evolved about ten times as fast as bivalves.

A second deduction follows from the difference in shape between curves for surviving and extinct genera. In bivalves, but not in carnivores, there are far more surviving genera with a very long survival than would be expected from the survival times of the extinct genera. A number of surviving genera have existed for over 300 million years. This means that there are a number of bivalve genera which are peculiar in that they have survived for a remarkably long time without undergoing any major evolutionary change. Groups of animals peculiar in this respect are not confined to the bivalves; other and perhaps more familiar examples of long-lasting groups include the lungfishes already described; the modern crocodiles, which differ little from crocodiles of the Jurassic period; the opossums of South America, which closely resemble the small pouched mammals which had a world-wide distribution in the late Cretaceous; *Sphenodon*, a lizard-like reptile now found only in New Zealand, but closely resembling an animal of Triassic age recently found in Gloucestershire by Lamplugh Robinson; and the coelocanth fish *Latimeria* recently caught off the African coast, but belonging to a group previously thought to have been extinct for some 80 million years.

Unfortunately one cannot give detailed reasons in any particular case for such long-continued survival with little evolutionary change. The lack of evolutionary change is fairly certainly not due to any peculiarities of the genetic system in these groups, rendering them incapable of change in response to selection. No studies have been made of the genetics or of the response to selection in any of these groups,* but the possibility that they have any inherent lack of evolutionary plasticity can be ruled out on other grounds. First, although all appear archaic today in that they resemble animals which lived a hundred million years or more ago, they were at the time of their origin advanced types; for example, the crocodiles belong to the group of 'ruling reptiles', the archosaurs, which were in their day the most advanced and successful of land vertebrates. Second, slowly evolving groups have often given rise through splitting to rapidly evolving lines; for example, animals very like opossums were the ancestors of the extremely varied mammalian fauna of Australia, and a genus of oysters, *Ostrea*, which

* See page 190 for a recent study of a 'conservative' animal, the Horseshoe Crab.

has undergone remarkably little change during long geological periods, has repeatedly given rise to rapidly evolving lines, originating independently, but given the same generic name, *Gryphea*, because of the similarity of the form evolved in various lines. Finally, slowly evolving groups show as great a variation within populations as do other groups, and have continuously undergone processes of speciation; the continuation of the 'group' means only that some of the species have retained its essential characteristics. In fact, the change in the genetic system which most seriously reduces evolutionary plasticity is the abandonment of sexual reproduction, and, as we have seen, this usually leads, not to long survival, but to early extinction.

In order to persist without evolutionary change, a group must be well adapted to a particular way of life in a particular kind of environment, a term which here includes both the physical features of its surroundings and the other animals and plants present. Changes in this environment must not be so great as to render impossible the way of life characteristic of the group. The continued survival, with or without change, of a group of organisms depends on the maintenance of a particular relationship between organism and environment. It is difficult, for any of the groups mentioned, to define precisely the nature of this relationship, but it was suggested above that lungfish are adapted to stagnant water liable to dry up, and similarly crocodiles are adapted to life in rivers, feeding mainly on land animals coming to the water to drink, and so on. The fact that they have survived for so long shows, not only that they are well adapted to this kind of life, but also that the circumstances in which such a way of life is possible have persisted. In the case of *Sphenodon* it seems that one necessary feature of the environment has been the absence of competition from mammals, since this reptile, recently widespread in New Zealand, and with a wide distribution in the Mesozoic, is now confined to a few offshore islands.

To sum up, such examples of very slow evolutionary change are to be explained, not by any inherent lack of evolutionary potentialities, but by the achievement of a stable relationship between organism and environment, capable of persisting over long periods. The more exciting problem, of particularly rapid evolutionary change, will be discussed in the last part of this chapter.

How well do the observed rates of evolution fit in with predictions based on genetic arguments? The simplest genetic change which can occur in a population is the replacement of one allele at a locus by another, owing to natural selection in favour of the new allele. Now selection implies the early death, or infertility, of individuals carrying the less favourable allele. Haldane has calculated the number of such deaths, spread over many generations, necessary before one allele has replaced another. He finds that, unless selection is very intense, the process usually involves a number of deaths equal to about ten or twenty times the number of individuals in the population at any one time, and occasionally to a hundred times this number; he suggests that the mean value is about 30 times the population number.

Now there is a limit to the number of selective deaths which can occur in a single generation, which will in part depend on the maximum number of offspring which a pair can produce. But even if a single female can lay millions of eggs, as can for example a cod or herring, the vast majority of deaths will be fortuitous, not selective; the few who survive will usually do so because they have been lucky, not because they have favourable genotypes. Even with intense selection, it seems unlikely that more than half the deaths will be selective, and Haldane suggests that an intensity of selection (see page 47) of 0·1 is a more probable figure. If so, it would take $30/0{\cdot}1 = 300$ generations of selection to replace one allele by another.

If changes are taking place at a number of loci at the same time, it is reasonable to assume the same *total* intensity of selection. That is to say, if a number of alleles $a, b, c \ldots$ are being gradually replaced by $A, B, C \ldots$, then a total intensity of selection of 0·1 means that in each generation one individual in ten fails to survive because it carries one or more of the alleles a, $b, c \ldots$ whereas it would have survived had it not carried those alleles. In this case, there will be an average of one gene substitution per 300 generations. Species probably differ by alleles at about 1,000 loci (that is, the number is probably greater than 100 and less than 10,000). If so, the evolution of a new species would take about 300,000 generations.

These calculations are only approximate. However, Zeuner has estimated that in mammals during the Pleistocene about 500,000 years were required for the evolution of new species; this agrees rather well with Haldane's estimate. But this estimate provides only an upper limit to the rate of evolution, which may often be much slower than this, as it has probably been for most bivalves during the last 300 million years. It may occasionally be more rapid, if the intensity of selection is greater than 0·1.

(ii) *Trends in Evolution*

In the investigation of fossils from an evolutionary viewpoint, one of the early successes was Kowalevsky's account in 1874 of the evolution of the horses. Kowalevsky was able to study a series of fossil genera from the Old World. These he arranged in a linear series according to their geological age, and this series showed successive increases in size, reduction of the digits to a single functional toe, changes in skull proportions, and increases in the height and complexity of the teeth. Later studies have confirmed that the kinds of changes described by Kowalevsky did in fact take place, but in other ways have greatly altered the picture which he presented. Six genera are now known from the Old World (*Hyracotherium*, *Palaeotherium*, *Anchitherium*, *Hypohippus*, *Hipparion*, and *Equus* itself), which, if arranged in order of their appearance in the fossil record, do show more or less progressive changes in structure. But there are sharp discontinuities in structure between these genera, which might suggest that evolution had taken place in a series of jumps. That this is not so has been revealed by the study of fossil horses in the New World. It seems that although the surviving horse species are confined to the Old World, the major evolution of the group took place in the New, and that the Old World genera represent successive immigrations from North America at times when a land bridge connected Alaska with Siberia. From the New World fossils it is possible to reconstruct in great detail the evolutionary history of horses. It is known that the picture based only on Old World fossils is false in two respects. First, the apparent discontinuities in structure are no longer present when the New World fossils are taken into account. Second, the history of horses must be represented, not as a single linear series, but as a much-branched tree with many independent lines of descent. A simplified diagram of this history is shown in Figure 28. Only genera are represented as independent branches; if species also were included the picture would be vastly more complex.

The history of horses has been studied in greater detail than that of any other group. It is therefore worth considering this history further, and in particular to see how far it can be explained as a series of adaptive responses to natural selection. Changes in size, in the skull and teeth, and in the limbs will be discussed in turn, although as will be seen the various changes are functionally related.

A. *Size.* Modern horses are considerably larger than their Eocene ancestor, *Hyracotherium*. There has in fact been a general tendency towards greater size in a number of different lineages, but this has by no means been universal. Three quite independent lines showed marked decreases in size, and the existing wild horses are slightly smaller than their ancestors in the Pleistocene.

None the less the commonest tendency has been towards greater size, and the same is true of other herbivorous groups, for example the cloven-hoofed mammals (Artiodactyls), the elephants (Proboscidea), and the herbivorous dinosaurs. The increase in size in some, but not all, carnivores can be explained as an adaptation to preying on the larger herbivores, but the causes of the increased size of the herbivores themselves are not obvious. It may in part be due to selection by predators. If at any given time the smaller individuals in a herd are more likely to be attacked, this will select in favour of increased size. Another possible cause (first pointed out by Watson, whose argument has been slightly modified here) concerns the food requirements and thermal efficiency of herbivores. An animal requires 'fuel' to keep its body working, to keep itself warm (in mammals and birds) and to move itself from place to place. It has been shown

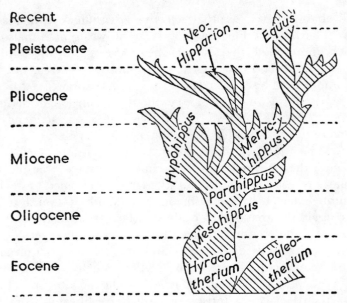

Figure 28. Simplified diagram of the genealogy of horses (after Simpson). Only the names of those genera mentioned in the text are given. The arrow indicates the lineage ancestral to the grazing horses.

that the energy required for these purposes is roughly proportional to the surface area and not to the volume of the animal: an animal twice the height requires four times as much food a day. Now one of the factors which determine how much food a herbivore can eat during its lifetime is the time it takes for its teeth to wear down to the gum. Now, since the volume of tooth worn away will be proportional to the volume of food eaten, an animal twice as large, with teeth eight times the volume, will, other things being equal, be able to eat eight times as much food before its teeth wear out. Since it needs only four times as much food a day to keep itself going, its teeth should last twice as long. Thus increased size should, without any other change, increase the potential life span of an individual. It is interesting that those herbivores, mainly rodents, which have remained small, and which seek to escape from predators by burrowing, have grinding teeth which continue to grow from the root throughout life.

Probably there are other selective advantages in increased size in herbivores. For example, large browsing mammals can reach more food than small ones. Also, since the rate of utilization of food is proportional to the surface area, but the volume of food reserves in the form of fat is proportional to the volume, a large animal can go without food for longer periods. Nevertheless, as the history of horses shows, the tendency to increase in size is neither inevitable nor irreversible.

B. *Skull and Teeth.* Early on in horse evolution, during the Eocene, the premolars were modified so as to resemble the molar teeth, so forming a continuous and lengthened battery of grinding cheek teeth, an easily understood adaptation. Further increase in the area of the grinding surface of the teeth could be achieved by increasing the length (parallel to the length of the jaw) of the individual teeth, and this in turn required a lengthening of the facial region of the skull to accommodate them. Such an increase in the length of the facial region of the skull is in fact characteristic of most horse lineages from the Oligocene onwards, leading to the familiar appearance of a horse's face. Robb has argued that the *relative* lengthening of the face may be a direct consequence of increased size. He has pointed out that larger horses tend to have relatively longer faces, and that this is true whether comparison is made between the same individual at different stages of growth, between large and small adults of the same species, or between large (usually recent) and small (usually primitive) species. This suggests that the developmental pattern of all horses is such that the face grows relatively more rapidly than the body as a whole, so that an increase in size results in a change in skull proportions. If so, selection for increased size would itself also bring about a lengthening of the face. It is, however, important to remember that the changes in skull shape were themselves adaptive, and that if they had not been so, it is probable that natural selection would have altered the developmental relationship between size and shape.

Another important change in the teeth has already been described (page 276); in those species which adopted the grazing habit, there was a rapid increase in the relative height of the grinding teeth. In these same lines, but not in the browsing horses, there also developed a more complex pattern of ridges on the surface of the teeth, of importance in chopping grass into short lengths before swallowing it.

C. *Limbs.* The most important changes in the limbs of horses concern the mechanism of the feet. Early horses had feet functionally similar to those of a dog; there were four functional toes on the fore feet and three on the hind, the weight being carried on soft pads. During the Oligocene one of the front toes was lost, but the padded condition remained. This type of foot was retained by all subsequent browsing horses. A dramatic change, well attested by fossils, occurred in the line *Parahippus-Merychippus*, which was ancestral to all later grazing horses. By a series of small, but, on an evolutionary time scale, rapid changes, a foot in which the padded digits lay flat on the ground evolved into one in which the weight when standing was carried on the tip of the central toe. When additional weight fell on the foot when galloping, the central toe was bent forward, stretching elastic ligaments in the foot. The subsequent elastic recoil of these ligaments then helped to spring the animal off the ground at the next stride. The lateral toes were retained, and acted as buffers or stops preventing the ligaments from being overstretched. Finally in *Pliohippus* there evolved the foot mechanism found in modern horses; the lateral toes are reduced to vestiges, and their functional role as stops has been taken over by check ligaments in the foot.

The interest in this story lies in the fact that the changes did not occur gradually and continuously. Three functionally distinct types of foot, the three-toed padded foot, the three-toed springing foot, and the one-toed springing foot, succeeded one another, each persisting for long periods with little change, and the transitions between them occurred rapidly in single lines of descent. Each type is probably an improvement, in a galloping mammal, on its predecessor. The rapidity of the transitions between them, however, suggests that intermediate types of foot mechanism would have a lower efficiency; consequently selection would act so as to preserve an existing mechanism, or, once some threshold had been passed, rapidly to perfect a new one.

The main trends in horse evolution can reasonably be explained as adaptations to a herbivorous life, produced by natural selection. When a change of habits occurred among the grazing horses, this was accompanied by rapid changes in the structure of teeth and feet. In so far as a number of related lineages tend to undergo similar evolutionary changes, this is because groups of animals which had adopted similar ways of life are subjected to similar selection pressures.

The direction of evolutionary change is determined not only by the environment, but also by the present structure and habits of the organism and the uses it makes of that environment. This is illustrated by the gaits of land vertebrates. The tetrapods are descended from fish which swam by lateral undulations of the body. Accordingly, all reptiles which run on four legs move their legs alternately, the right hind leg and left fore leg being forward when the left hind leg and right fore leg are back, and vice versa; these limb movements are assisted by sideways flexions of the body similar to those of fish. It follows that those reptiles which have achieved high speed by adopting a bipedal gait are runners; they move their hind legs alternately. Of these the most important were the extinct archosaurs, including the bipedal dinosaurs, and their descendants the birds. All birds which travel fast on the ground run; only some of the perching birds hop. In contrast, all bipedal mammals except man progress in a series of leaps (e.g. kangaroos, jerboas, elephant shrews, Cape jumping hares). All these forms have evolved from relatively small galloping mammals; the gallop is a gait unique to mammals, in which the hind legs tend to move together and in which the backbone is flexed in a vertical and not a lateral plane. The exception among bipedal mammals is man, descended from arboreal and not from galloping ancestors. This contrast between bipedal mammals and reptiles shows how the direction of evolutionary change depends on the structure and habits of the ancestral population.

When a reversal or change in the direction of evolution has occurred, it usually reflects a change in the environment in which the population lives, or perhaps more often a change in the methods of exploiting that environment. One of the most

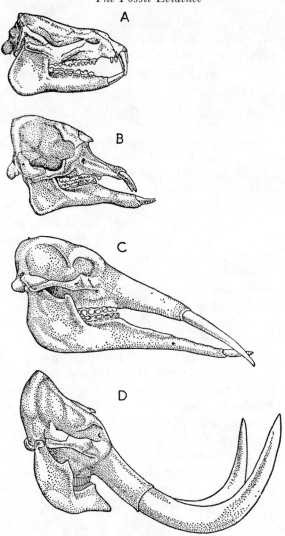

Figure 29. Skulls of A, *Moeritherium*; B, *Phiomia*; C, *Trilophodon*; D, *Mammonteus*, the woolly mammoth (after Romer).

entertaining illustrations of reversal in evolution is afforded by the Proboscidea (elephants), and concerns the evolution of that unique structure, the elephant's trunk. Since the trunk contains no bone, direct evidence of its evolution is lacking, but its course and causes can be deduced from changes in the bony skull. Our knowledge of proboscidean evolution depends largely on the work of Andrews, amplified by Osborn; the functional interpretation given below of changes in the skull and teeth was suggested by Watson.

The earliest known genus, *Moeritherium*, from the upper Eocene, although probably not directly ancestral to later forms, must closely have resembled those ancestors. It was a small animal, standing about two feet at the shoulder. To a still greater extent than in horse evolution, most subsequent lineages have increased in size. Probably the most important cause of this increase has been that elephants rely for safety from carnivores on large size and defensive weapons rather than on flight. The limbs have changed in adaptation to supporting a greater weight. They have become relatively as well as absolutely stouter, and are arranged in straight columns (as are the legs of men but not of most other mammals), so that they are exposed to direct compression stresses rather than to the more severe bending stresses imposed on the limbs of smaller mammals which stand with their limbs partly flexed.

The lower jaw of *Moeritherium* was heavy, and the lower incisors projected forwards, and were probably used in grubbing for food on the ground, something in the manner of a pig. This method of feeding was retained in later Proboscidea (see Figure 29). There was a progressive and extreme lengthening of the lower jaw, carrying the lower incisors, sometimes modified to form a perfect shovel, forward of the face. These incisors worked against a horny pad carried on a fleshy and probably mobile extension of the face, the forerunner of the trunk. The wear on these incisors shows that they were still used to dig or shovel up food; this food could no longer be conveyed to the mouth by the lips, so this job was taken over by the fleshy extension of the face, with the probable result that this new organ increased in mobility. At this stage the upper incisors played little or no part in feeding, and early acquired the function of defensive tusks.

In time a stage must have been reached at which the mobility of the trunk was such that it became an efficient organ in the gathering of food, without assistance from the lower jaw and

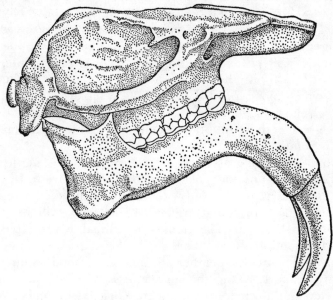

Figure 30. Skull of *Dinotherium* (after Romer).

incisors. From then on, the enormously long lower jaw was a
hindrance rather than a help. In one group (e.g. *Dinotherium*,
Figure 30) the lower jaw and incisors were bent downwards and
backwards to carry them out of the way of the freely hanging
trunk. In other lines there was a progressive and rapid
shortening of the lower jaw, and a reduction and final loss of the
lower incisors, until the condition in modern elephants was
reached. Thus the trunk, originally evolved to act in unison
with the lower jaw, finally reached a threshold at which it was
by itself an effective instrument for gathering food, and this was
followed by a reversal in the direction of evolution of the lower
jaw.

In the evolution of their teeth the Proboscidea provide both
parallels and contrasts to the horses. The early forms probably
fed on roots, twigs, and leaves, which they crushed between low-
crowned teeth, in which knobs on one tooth fit into valleys in
another. But, as in the horses, one group, which includes the

surviving elephants and recently extinct mammoths, but not the mastodons, adopted the habit of feeding partly or wholly on grass. The problem of tooth wear has, however, been solved in a different way. In an elephant, only four molar teeth, one on each side of each jaw, are functional at any one age and time. As these teeth wear out, they are replaced by the next molars in the series, which move forward in the jaw to replace them; thus, although the total number of teeth developed during a lifetime is no greater than in other mammals, tooth replacement continues into old age.

Before digestion, grass must be cut up into short lengths by the scissor action of ridges of enamel on opposing teeth. In two animals of different sizes, but with geometrically similar teeth, the total length of enamel ridges on the teeth will be proportional to the linear dimensions of the animal (for example, to its height), and so also will be the quantity of grass it can chew in a day. As pointed out earlier, the quantity of food required is roughly proportional to the surface area. Consequently in a large elephant, to chew up enough grass in a day becomes a real problem, particularly since the methods of tooth replacement described above mean that the total surface area of the teeth in action at any one time is, relative to the size of the animal, rather small. This difficulty has been partly overcome by a great increase in the complexity of the pattern of enamel, which in modern elephants forms numerous series of ridges across the teeth. Despite this, however, an elephant may have to spend up to eighteen hours a day eating.

(iii) *Extinction*

It is the ultimate fate of most species to become extinct. Romer has estimated that perhaps not more than one per cent of the land vertebrates of the middle Mesozoic have left living descendants. It may take much less time for a species to disappear by extinction than by evolutionary transformation. It is therefore sad but true that we have had far more opportunities to observe the former process.

Probably all recent cases of extinction can be ascribed directly or indirectly to man's activities. Occasionally the cause has been direct predation by man; the flightless dodos and great auks disappeared because they were killed or their eggs taken by men. But more commonly man's influence has been less direct. Agricultural practices may destroy the habitats natural to a particular species; for example, the Bearded Tit has been drastically reduced in numbers by the draining of the fens. Of all causes of extinction, the one which will in future be most difficult to control is the introduction of animals and plants into areas previously closed to them, with the consequent competitive elimination of the native flora or fauna. For example, the native British Red Squirrel, although holding its own in some areas, is now extinct in much of southern England owing to competition from the introduced American Grey Squirrel. The giant tortoises of the Galapagos Islands have been greatly reduced in numbers by competition from goats originally landed on the islands by Captain Cook as an insurance against shipwreck. It should not be thought that competitive elimination of this kind implies that the invading species attacks or kills its rivals; goats do not eat tortoises, they merely eat the plants that tortoises eat.

Similar effects have been produced in the past without human intervention, from changes in climate or geography, or the evolution or immigration of new predators or competitors. In the fossil record, however, it is more often possible to trace the extinction of a group of related species than of a single species. It is easy enough to state the general truth that changes in the environment, and in particular in the other animals and plants present, made the way of life characteristic of such groups no longer possible. It is much more difficult to give more precise reasons in any particular case, since we can never have a complete picture of the requirements which any extinct group made of its surroundings.

Sometimes the extinction of a group can be explained by competition from a more recently evolving, or invading, group. An example of this kind is the extinction of the South American herbivores after the establishment of a land bridge between North and South America. Another example is given by the work of Jepsen on the extinction of the multituberculates, a group of mammals appearing in the Jurassic (i.e. before placental or marsupial mammals) and disappearing in the Eocene. These creatures closely resembled rodents in their adaptations for gnawing (Figure 31). The latter, however, are certainly not descended from the former. Jepsen has recorded the variety of rodents and of multituberculates found in a series of deposits in the Rocky Mountain region of North America; his results were as follows:

		Multituberculates		Rodents	
		Genera	Species	Genera	Species
Eocene	late	0	0	13	31
	middle	0	0	9	19
	early A	0	0	3	8
	early B	3	5	1	4
Palaeocene	late	7	11	1	1
	middle	6	17	0	0
	early	5	7	0	0

It is difficult to avoid the conclusion that the multituberculates died out because they could not compete with the more efficient rodents.

Unfortunately, it is seldom possible to provide so convincing an explanation. This is particularly true of the most dramatic of all extinctions, that of the 'ruling reptiles', the archosaurs, which were the dominant land vertebrates for most of the Mesozoic. These archosaurs, with the exception of the crocodiles and of the birds (the latter are their descendants), disappeared towards the end of the Cretaceous. The extinction of the pterodactyls may have been due to competition from birds, but the disappearance of the terrestrial archosaurs is less easy to explain, since at the time of their disappearance no large or medium-sized mammals of similar adaptive type existed. Many ingenious suggestions have been made, but none are fully satisfactory. What is clear, however, is that the extinction of the archosaurs, whatever its cause, did open the way for the subsequent adaptive radiation of the mammals.

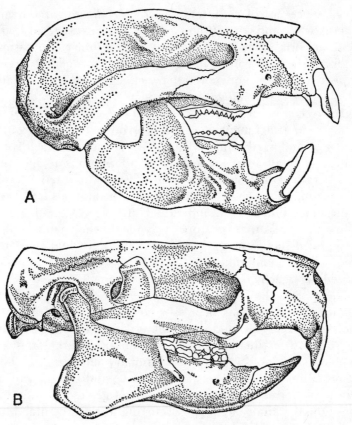

Figure 31. Skulls of A, *Taeniolabis*, a multituberculate; B, *Paramys*, an Eocene rodent (after Romer).

There are some cases of extinction which have been interpreted as a result, not of changes in external conditions, but of the continuation of an evolutionary trend beyond the point at which it is adaptive, as a kind of racial suicide. Such an interpretation is difficult to accept, since we know of no mechanisms which would cause such an inadaptive continuation of an evolutionary trend. Therefore the classic case of this supposed phenomenon will be described, to see whether it can be explained in any other way.

During the Mesozoic there arose from an ancestral oyster

stock a succession of evolutionary lines, all changing in the same general way, and all ultimately becoming extinct. Although such lines originated independently on a number of occasions, the forms which evolved are given a common generic name, *Gryphea*. The course of evolutionary change is shown in Figure 32. The relevant changes are an increase in size and in the degree of coiling of the lower valve of the shell, and a change from a rigid attachment to a solid substrate (A, B, C) to a form resting on and partly embedded in a soft substrate. The first two changes can also be understood as adaptations to life on a muddy rather than on a rocky bottom, since both tend to raise the opening of the shell, through which the animals feed, clear of the mud. But in some individuals in stage D the degree of coiling is so great that the upper valve of the shell can no longer open; once such a stage is reached the individual must die. Frequently, populations of *Gryphea*, in which a number of individuals reached this stage, became extinct.

These facts have been interpreted as showing that an evolutionary trend has been carried, by some kind of 'evolutionary inertia', beyond the point at which it is adaptive, and has so led to extinction. But an explanation of the facts, not involving any non-adaptive evolutionary changes, has been suggested by Westoll. The stages A, B, C, and D represent not only adult forms at successive periods of geological time, but also show the kinds of change occurring during the individual growth of the latest forms (except that in individual growth no change took place in the method of attachment). Thus an individual which in old age died because its shell would not open might nevertheless be better adapted earlier in life than were the less coiled individuals. There would therefore be a balance between selection in favour of a high degree of coiling when young and against it in old age. Now most individuals in most populations die anyway before reaching old age, perhaps because they are killed and eaten. Therefore such a balance of selection will tend to produce advantageous changes in juveniles at the expense of damaging or lethal changes in old age, because, if few individuals survive into old age, selection in favour of characteristics which are advantageous only in old age

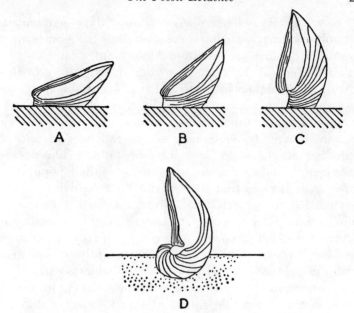

Figure 32. The evolution of *Gryphea* (after Westoll)

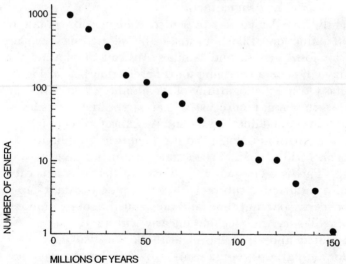

Figure 33. Survivorship curve for genera of bony fishes (after Van Valen).

will be relatively ineffective. As Medawar has suggested, it is
probably for this reason that there is a decline in vigour with age
in most individuals of most animal and plant species.

Thus, once the changes occurring during the growth of
individuals are taken into account, the evolutionary changes
occurring in the various *Gryphea* lineages, including the occurrence of many individuals so coiled as to be unable to open
their shells, can be explained as consequences of selection
improving adaptation to life on a muddy bottom. This does not
by itself explain the subsequent extinction of these populations.
However, it is clear that their particular adaptive trend had
been carried to a mechanical limit; any further increase in
coiling would indeed result in racial suicide as soon as the final
closure of the shell occurred before the achievement of sexual
maturity. With no possibility of further adaptive change, such
Gryphea populations would be peculiarly liable to extinction if
the environment changed; we do not know what environmental
changes were responsible, but it is easier to imagine that such
changes took place than that an inherent evolutionary urge
drove *Gryphea* to its own extinction.

Recently Van Valen has presented evidence for a 'law of
constant extinction'. Like Simpson, he has plotted survivorship
curves for fossil genera and families. Instead of plotting the
percentage of genera surviving for different times, as in Figure
27, he has plotted the logarithm of the number of survivors; an
example is shown in Figure 33. As is apparent, the points lie on
a straight line. What does this mean? We expect to get a straight
line on a logarithmic plot if the probability per million years
that a genus will go extinct is constant in time for all genera in
a group. This is analogous to radioactive decay, when each
atom of a radioactive substance has the same constant probability of decay per unit time, and the logarithm of the number
of atoms still surviving declines linearly with time.

Van Valen finds that the logarithmic survival curves are
linear for most groups of organisms, provided that the members
of the group have something in common ecologically as well as
taxonomically; in the example given, all bony fish live in water
as well as being related to one another. The slope of the line

differs for different groups, implying different extinction probabilities. There are of course exceptions to the rule. One type of
exception is the occurrence of genera with abnormally long
survival times, as discussed on pages 281–4. Another type of
exception occurs when all or almost all members of a group
became extinct at a particular geological period. Despite these
exceptions, however, the rule holds in a great majority of cases.

The explanation is not clear. Van Valen's own suggestion,
which is probably along the right lines, is as follows. The
members of a taxonomic group compete for limited resources.
Hence each evolutionary advance by one species is experienced
as a deterioration of the environment by the others. The various
species are therefore engaged in a kind of evolutionary race, in
which the losers go extinct. Van Valen calls this the 'Red
Queen' hypothesis; the Red Queen, you will remember,
explained to Alice that it takes all the running you can do to stay
in the same place. Unfortunately it is not clear, at least to me,
why a constant rate of extinction should necessarily follow from
this idea of a continuous evolutionary race.

(iv) *The Origins of Major Groups*

In the previous discussion I have used the terms 'lineage' or
'line of descent' without bothering to define them, but in the
hope that their meaning would be sufficiently clear. It is now
necessary to examine the facts behind these terms in a little more
detail. In Figure 34, copied from Simpson, the patterns of
descent in sexually reproducing organisms are shown diagrammatically, on three different scales, and greatly simplified in
order to obtain a two-dimensional representation. In A the
ascending lines represent individual lives, and the cross bars
sexual reproduction, the whole representing individual descent
within a single population. B represents descent in a single
species. Each line represents a single population, at times
isolated but occasionally merging with other populations. Each
such line has a structure as shown in A. In C the various lines
represent lineages; each has a structure as shown in B. The lines
branch, representing the division of a single species into two, but

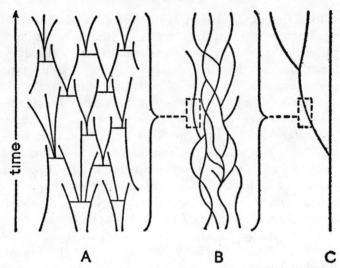

Figure 34. The patterns of descent in sexually reproducing organisms (after Simpson). For explanation see text.

never rejoin. We have seen in the last chapter that hybridization between species does occur and may have important consequences. But such hybridization takes place only between closely related species; lineages which have been separate for a reasonably long time do not merge. The occurrence of interspecific hybridization means that the distinction between patterns of types B and C is not a sharp one, but it is still true that evolutionary events on a sufficiently long time scale can be represented by a diagram of type C.

By a major group of animals or plants we mean a large group of species all of which have certain characteristics in common, as do, for example, all mammals or all insects. Now if the lineages of all existing mammals could be traced back in time, they would ultimately converge into a single lineage, although, as we shall see later, it does not follow that the individuals in this ancestral line possessed all the characteristics of modern mammals. It is, however, the events occurring in such an ancestral lineage and its immediate ancestors and descendants which are referred to by the term 'the origin of a major group'.

It follows that one can only judge retrospectively whether the changes occurring in a particular lineage contribute to the origin of a major group; they do so if the descendants of that lineage, by speciation and adaptive radiation, give rise to a large and varied group of animals. Thus the characteristic feature of a lineage ancestral to a major group is that there evolved within that lineage some character or group of characters which are of selective value for animals living in a variety of different ways in a wide range of environments. To use an engineering analogy, an 'invention' is made which subsequently proves to have a wide range of applications. An obvious example is afforded by the origin of the birds from the archosaurs. We are fortunate in having two fossils, *Archaeopteryx* and *Archaeornis*, intermediate between the archosaurs and modern birds. As far as their skeletons are concerned, these two animals are typical small bipedal archosaurs; if nothing else were known of them they would not easily be recognized as bird ancestors. But luckily impressions of their feathers can be made out. It was the evolution of feathers in this group which ensured that it would leave such a large and varied assembly of descendant species.

Feathers can perform at least three major functions. Not only do they provide a light and rigid aerofoil section for the wings, but they also trap a layer of air over the surface of the body, which enables birds like gulls and ducks to swim on the surface of water, and, of more general importance, which helps to keep the body warm.

This example will help to explain one of the difficulties often encountered in explaining evolution in terms of natural selection. It often seems that a perfected organ, although efficient at performing its function, is far too complex to have arisen by one or a few mutations, and yet is such that any intermediate stage between the absence of the organ and its full development would be incapable of performing this function. Thus it is inconceivable that the flight feathers of a bird could have arisen by a single mutation, but the intermediate stages between a scale and a feather would be useless for flight. In this case the difficulty disappears once it is realized that during the early stages of the evolution of feathers, the latter were probably of selective advantage because they conserved heat, and only later did they become functional in flight.

This is a very common feature of evolution; a new structure evolves at first because it confers advantage by performing one function, but in time it reaches a threshold beyond which it can effectively perform a different function. We saw earlier that something of this kind occurred during the evolution of the elephant's trunk. The flying membranes of bats and of pterodactyls were probably used in gliding before they were of any use in flapping flight, and, as Spurway has pointed out, small membranes along the sides of the body are found in some arboreal mammals which do not even glide, and these folds of skin render such animals more difficult to see by eliminating the shadows they would otherwise cast. Similarly, lungs were a selective advantage to fish living in stagnant waters, enabling them to breathe air, long before the descendants of these fish walked on land; in modern teleost fishes the lung has lost its function as a breathing organ, and has been transformed into a hydrostatic organ, the swim bladder. These examples show that there is no reason to suppose that even the most complex structures underwent a long period of evolution and elaboration before they could function, and so confer selective advantage; rather their function may have changed once or even several times in the course of evolution.

The evolution of feathers was the decisive event in the origin of birds. This achievement was followed by many other changes improving the powers of flight, of which the most striking were the development of a great keel on the breast bone to support the flight muscles, and the reduction of the long feathered tail to a short fan.

The birds are characterized by this one unique feature, the possession of feathers, and it is very likely that these evolved once only in a single lineage. The story is different for mammals, which differ from reptiles in a whole complex of characters, of which the following are among the most important:

(a) Their method of locomotion. The elbows have been rotated backward and the knees forward so as to lie under the body, and the characteristic gait is the gallop, in which the legs are moved and the backbone is arched in a vertical plane, and not in a horizontal plane as in a lizard.

(b) Their teeth are replaced only once, and there is a 'division of labour' between incisors, canines, and molar teeth.

(c) The air passages from the nose are separated by a secondary bony palate from the food passage from the mouth, thus enabling an animal to breathe and chew at the same time.

(d) They have lost the original bony covering of the sides of the head, retaining only the cheek bone; consequently there is more room for the free play of the temporal muscles working the jaws.

(e) The lower jaw is formed from a single bone, the dentary, articulating directly with the side of the skull, while the two bones (quadrate and articular) which in reptiles form the jaw articulation, in mammals help to conduct sound impulses from the eardrum to the inner ear.

(f) They are warm-blooded, and covered with hair.

(g) They bring forth their young alive, the mother nourishing the unborn foetus through a placenta.

(h) After birth the young are fed on milk.

These various features did not all evolve at the same rate or time; in fact the living monotremes (*Platypus* and *Echidna*) still lay eggs and are primitive in some other respects. One or other feature may have been lost in some lineages; for example, bats are torpid when at rest, and whales do not gallop, although they still retain the arching movements of the backbone characteristic of small galloping mammals.

Little has been deduced from the fossil record about the evolution of the last three characteristics, but a good deal is known about the first five. One of the first great groups of reptiles to undergo an adaptive radiation were the 'mammal-like reptiles', which were the dominant land animals of the Permian, until they were eclipsed, but not wholly extinguished, by the archosaurs. Progressive changes in all these five features can be followed among these animals, from structures typical of primitive reptiles towards those typical of mammals. It has been found desirable to pick on some one characteristic to define a mammal, and the choice has settled on feature (e) above; to an anatomist, a mammal is an animal with a single bone in its lower jaw, just as a bird is an animal with feathers. But whereas the latter definition is natural as well as convenient, the former has little to recommend it but convenience. Changes in all these features, not only in one, were important in the origin of the mammals. In fact features (c) and (d) were effectively complete, and (a) and (b) well on their way, in animals which by the accepted definition were still reptiles. Further, it seems possible that the characteristic mammalian jaw and ear bones evolved not once but in several lineages. If so, it follows that if the lines of descent of all mammals, living and extinct, could be traced back until they converged in a single lineage, then that lineage would consist of animals which, on our present criteria, were not themselves mammals.

This kind of difficulty in naming is common in palaeontology; it does not in the least matter, so long as the situation is understood, whether an animal is called a mammal or not. It is helpful to choose one diagnostic character to define a group, provided this does not lead one to suppose that all the other characteristic features of that group originated simultaneously. In fact the ancestors of the mammals can be distinguished even among the earliest known reptiles, and their evolution proceeded throughout the Mesozoic, although the great adaptive radiations of the mammals were delayed until the stage was set by the extinction of the archosaurs.

None the less it is true of the first mammals, as of the first birds, that they were a far less abundant and varied group than their living descendants. It is probably always true of the origin of a major group that one or more of the decisive changes involved took place in a single lineage, that is to say in a single species probably confined to a restricted part of the world. Further, such decisive changes probably occurred because a population was exposed to new environmental conditions, or, which is only an active rather than a passive way of expressing the same thing, because it adopted new habits. In either case it would be exposed to new selective forces, and would be likely to evolve particularly rapidly. For example, when the feathers of birds first reached the threshold at which they were useful in gliding as well as in heat conservation, there would be both ample room for improvement and strong selection in favour of it. Rapid evolutionary change would be expected, until a stage was reached beyond which no further improvement along similar lines was possible. In fact the evidence of *Archaeopteryx* suggests that the flight feathers had already reached a fair degree of perfection, although the skeleton was but little modified from the archosaurian condition.

Now if it is true that decisive evolutionary advances would be expected to take place by rapid evolution in single species (or at most, groups of related species) confined to a particular part of the world, it follows that the number of individuals representing any particular structural stage is very small when compared to the number of individuals at a given stage in a larger group of animals evolving more slowly. Consequently, transitional forms are less likely to be found as fossils. It is, in fact, the case that major groups often appear suddenly in the fossil record, and although it is usually possible to identify the group from which they have originated, intermediates are rare; sometimes, as in the case of *Archaeopteryx*, one is lucky. Strictly, the rareness of such intermediates is a confirmation of the view that the origin of major groups occurs rapidly in a limited population, rather than a deduction from it.

It is now possible to perceive a pattern of evolution often characteristic of the origin of a major group. At their first appearance, new structures or organs often evolve because they perform a function different from that which they will serve when fully elaborated. A threshold is reached, beyond which organs can acquire new functions, perhaps enabling their possessors to adopt new habits or colonize new habitats. Such an evolutionary breakthrough is achieved by a single species, or perhaps by a group of related species. Once it has occurred, natural selection will cause rapid improvement and elaboration of the new structures. This is followed or accompanied by speciation and adaptive radiation, whose extent is determined by the range of habitats and ways of life in which the new pattern of structural organization can confer a selective advantage over more primitive competitors.

One consequence of this pattern of evolution is the replacement in time of one group of animals with a common structural plan but with a wide range of adaptive types, by another group more advanced in general structure but with a similar range of adaptations. For example, during the Eocene there existed a group of carnivorous mammals, the creodonts, which in size and in tooth structure included species resembling modern bears, weasels, hyenas, and even sabre-tooths. But modern carnivores of these various kinds are not each descended from creodont species of similar adaptive type to themselves. Instead, the modern land carnivores arose from a single family (and possibly from a single genus or even a single species) of creodonts, whose descendants have subsequently radiated, and replaced the more archaic creodonts. The modern carnivores have, relative to their size, much larger brains than the creodonts, and it may be that their greater intelligence, of value to any kind of carnivore, was the main cause of their success.

A parallel was drawn above between this type of process and an engineering invention of wide application. To pursue this analogy further, it is worth remembering that such an invention is usually made to perform some limited task. For example, the steam engine was originally developed to pump water out of mines, and it has been suggested that printing was first developed by the Chinese for the multiplication of Buddhist formulas for devotional purposes; both inventions have since proved to be of general utility. The same is true of many biological 'inventions'. Even the wholly different patterns of organization characteristic of the major phyla first evolved to perform some relatively limited function. For example, a characteristic feature of the chordates is their possession of an elastic axial rod, the notochord (or its replacement, a jointed vertebral column), flanked by segmental muscles, the myotomes. This basic structural plan first evolved to make possible the mode of swimming by throwing the body into lateral sinusoidal waves, still used by primitive vertebrates like the lampreys, and, in a modified form, by many modern fishes. But this structure has since proved effective in land animals also. The variation observed today in the various phyla gives a measure of the range of applications possible to a structural pattern first evolved to perform some particular function in a small and relatively uniform group of animals.

In technology, old-fashioned ways of doing things may persist alongside the new, albeit in a restricted field. A horse and cart is more efficient than a van with a petrol engine for delivering milk, although it seems that the horse may now be replaced by the electric van. Similarly in the organic world the old may persist alongside the new; flatworms may resemble in structure the ancestors of most of the other phyla, but they still survive in the modern world. The analogy must not be pressed too far. In the present context the most important difference between technological and organic evolution is brought out in Figure 34 C. In the organic world, once two lineages have diverged for some time, they cannot rejoin. In engineering, two inventions, first developed to perform different functions in different kinds of machine, can be brought together in a single machine; the trolley-bus is a 'hybrid' between a bus and a tram. It is because of this restriction on the possible patterns of organic evolution that it is so common to find that a structure evolved to perform some new function arises as a modification of an already existing structure performing some different function. The evolutionary future of a group is determined not only by the environmental conditions which it meets, but also by its past history and present potentialities.

CHAPTER 18

Evolution and Development

As a rule, only adult animals are found as fossils. Embryos and larval forms are small, and seldom have hard parts which will fossilize. Sometimes, particularly in molluscs, it is possible to deduce a good deal about the appearance of young individuals from the structure of adults, who carry around with them, a permanent record of their shape when young. But it is in general true that the fossil record is a record of adult forms. Consequently the 'phylogeny' which can be deduced from fossils consists of a series of adults at successive stages in a line of descent. Each adult individual, however, was the end-product of a process of development; the development of an individual from fertilized egg to adult is called its 'ontogeny'. Thus although evolutionary changes are usually described in terms of the differences between successive adults, i.e. as phylogenetic changes, the differences between those adults were the consequence of differences between the paths of development which gave rise to them, i.e. of ontogenetic changes; phylogenetic changes are the result of changes in ontogeny. It follows that a study of ontogeny, even though confined to living animals, can throw a good deal of light on the processes which in the past were responsible for phylogenetic change.

The task of describing accurately the ontogenies of animals was effectively started by von Baer. From his studies he drew some general conclusions. In particular, he noted that the young stages of different animals often resembled one another more closely than did the respective adults. Later in development, an animal tends to depart more and more from the form of other animals. The early stages in the development of an

310

animal do not resemble the adult forms of other animals, but tend to resemble the early stages of those animals. As we shall see, there is much truth in these observations. A classic example is the resemblance between the early stages in the development of a bird and a mammal.

After the general acceptance of evolutionary views by biologists, the resemblances between the ontogenies of different animals was given a new interpretation in Haeckel's 'law of recapitulation'. According to this view, an individual during its ontogeny goes through a series of forms resembling its adult ancestors; ontogeny recapitulates phylogeny. This has been expressed in the phrase 'each animal during its development climbs up its family tree'. An example of the kind of evidence on which this view was based is the presence in the embryos of both birds and mammals of a series of gill pouches, not present in adults, although gill slits were present in the adult fish from which both birds and mammals are descended. Now if Haeckel's theory is correct, the relevance of a study of ontogeny to evolution is obvious, since it should be possible to discover by studying the development of an individual the kind of adult ancestors it had.

Unfortunately the facts do not support Haeckel, but suggest that von Baer was right in suggesting that young stages resemble, not adult ancestors, but the young stages of those ancestors. For example, the gill pouches of an embryo mammal are much more like the gill pouches of an embryo fish than they are like the gill slits of an adult fish. Still clearer evidence is afforded by the presence in embryos of structures which could not possibly have existed, in however modified a form, in any adult ancestor. Such are the extra-embryonic membranes, the amnion, allantois, and yolk sac, of developing birds, reptiles, and mammals. The evolution of the amnion and allantois was one of the decisive steps in the conquest of dry land by the vertebrates, since they serve to protect an embryo enclosed in a shelled egg on dry land, and enable it to respire. The allantois in reptiles and birds is a respiratory organ absorbing oxygen which has diffused through the shell; in placental mammals a derivative of the allantois contributes to the embryonic component of the placenta. These membranes therefore are embryonic adaptations. The resemblances between the extra-embryonic membranes of birds and mammals suggest that both groups are descended from a common reptilian ancestor which itself possessed such membranes as an embryo, but which certainly could not have possessed them as an adult. These facts were known to Haeckel, who recognized that they constituted an exception to his rule of recapitulation; but as more facts have accumulated it has become apparent that such exceptions are so common that the rule itself must be abandoned.

The retention in evolution of the extra-embryonic membranes is explained by the fact that they have specific physiological functions to perform; without them, the embryo could not live, any more than an adult could live without lungs. But the embryos of birds and mammals resemble one another in the presence of a number of other structures which have no obvious physiological function to perform. For example, embryo birds have a series of gill slits, and embryo mammals have gill pouches, similar in position to gill slits, but differing in that they end blindly and do not open at the surface of the embryo. Now these structures certainly do not act as respiratory organs, as do the gill slits of adult fish. But their presence cannot be explained merely by using the word 'recapitulation'. Even though they may not be necessary for the immediate survival of the embryo, it seems likely that in some way they contribute to the proper development of an adult; they have a developmental rather than a physiological function to perform. It is far from clear what this function is. It is known that the first gill pouch becomes the eustachian tube connecting the middle ear to the throat, and that its outer blind end becomes the eardrum; the other pouches help to form the tonsils and the thymus and parathyroid glands. But these facts seem insufficient to account for the continued presence of a series of gill pouches in embryo mammals, and probably they have other functions to perform, possibly connected with the development of the arteries which pass between them. Whatever the truth may prove to be, the point of this example is to suggest that structures which have lost their original physiological functions may persist in embryonic life because they play a part in the long chain of causes and effects which make up an individual development.

Any evolutionary change in adult structure is the consequence of changes in development in successive generations. It is not always true that change occurs by adding on additional chapters at the end of development. In fact, some of the most dramatic events in evolution have been the consequence of an opposite type of change, in which sexual maturity is achieved at an earlier developmental stage, so that previously embryonic characters are retained in the adult, and previously adult characters lost altogether. This process is known as neoteny.

A classic example is provided by the axolotl, a relative of newts and salamanders. Most newts have an aquatic tadpole stage, with gill slits and external gills, which, by a process of metamorphosis, is transformed into a mainly terrestrial lung-breathing adult. The laboratory axolotl normally becomes sexually mature without metamorphosis, retaining its external gills. It can, however, be induced to metamorphose by feeding thyroxin. The wild species from which the laboratory axolotl is probably derived normally undergoes metamorphosis. There are, however, neotenic populations in certain lakes in Mexico, but these differ in colour and in other ways from the laboratory axolotl, and should perhaps be regarded as belonging to a different species. The main interest of this example of neoteny lies in the fact that a number of species of newts, for example the 'mud puppy' *Necturus* and the blind cave newt *Proteus*, never metamorphose, and cannot be induced to do so experimentally.

Neoteny has played an important part in the evolution of man, as can be seen by comparison with his living relatives, the great apes. It must, of course, be remembered that man cannot be descended from contemporary apes, and that in fact since the lineages of man and the apes diverged each has undergone further evolutionary changes, adaptive to life on the ground and in the trees respectively. However, at least as far as the characteristics discussed below are concerned, it is fairly certain that the common ancestor of man and the apes resembled the latter rather than the former. The reason for regarding man as neotenous is that, as was first pointed out by Bolk, he in many ways resembles a young ape more closely than he does an adult ape, and therefore, by implication, that he resembles the young stages of his ancestors more closely than he does the adult stages of those ancestors. It is worth emphasizing that this is the precise opposite of the situation envisaged in Haeckel's theory of recapitulation, whereby the young of existing forms should resemble their adult ancestors.

Some respects in which a man resembles a young rather than an adult ape are as follows:

(a) In the shape of his skull. In most mammals the backbone is held roughly horizontal, and the eyes look straight forwards or sideways. In man, although the eyes look forwards, the backbone is held vertical, a condition which requires a 90° flexure of the skull. It is shown in Figure 35 that such a flexure exists in the skull of an embryo dog (and in fact in almost all vertebrate embryos) but is lost in the adult; in man the embryonic condition is retained. In this respect apes are intermediate between men and dogs. Other embryonic features of the human skull include the late closure of the sutures between the bones of the skull roof, the flat face, and the relatively large braincase.

(b) In his hairlessness. The soft covering of hair of newborn human babies corresponds to the hair of embryo apes which is lost before birth.

(c) In the relative proportions of his limbs and trunk.

These features are sufficient to indicate that at least some specifically human characteristics have been evolved by re-taining embryonic characters in the adult. The importance of other changes in relative rates of development in man will be discussed from a different point of view in the last chapter.

One possible reason for the importance of neoteny in the origin of major groups is as follows. Such groups are charac-terized, among other things, by their methods of locomotion. This is true, for example, of annelids, vertebrates, arthropods, molluscs, and echinoderms. In fact the names of the first three of these phyla refer to structures whose main function is in locomotion; the ringed appearance of annelid worms reflects

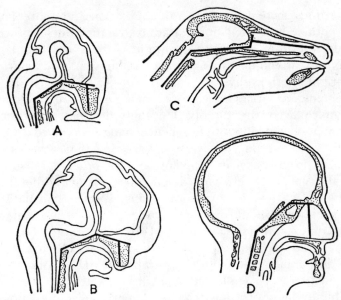

Figure 35. A series of sections showing the angle which the head makes with the trunk in A, an embryo dog; B, an embryo human being; C, an adult dog; D, an adult human being (after Bolk).

the segmentation of their bodies, including their muscles; the vertebral column first evolved as an aid in the characteristic undulating swimming movements of primitive aquatic vertebrates; the arthropods are animals with jointed limbs. A characteristic organ in molluscs is the muscular 'foot', which in bivalves may be used in burrowing, in gastropods forms a flat creeping sole, and in cephalopods is extended into series of arms surrounding the mouth and used in capturing prey, the main method of swimming being by jet propulsion. The echinoderms possess a unique system of tubular feet. These facts suggest that the evolution of a characteristic method of locomotion by a particular group was partly responsible for its subsequent success; it was the invention which proved to have a wide field of application. Now many marine animals are either sessile or bottom-living as adults, but have free-swimming larvae important in dispersing the population and in colonizing new habitats. Thus it may often be the case that locomotor adaptations arise first as larval adaptations, and are later incorporated by neoteny into the adult structure. Such a process is unlikely to be relevant to the origin of the mainly sessile or bottom-living echinoderms, but may well have played a part in the origin of other phyla.

Neoteny is a process whereby a striking evolutionary change may take place rather rapidly. This raises the question: is there anything special about the genetic changes involved in the origin of major groups? If we accept the view that varieties are incipient species, species incipient families, orders, and so on, then the types of genetic variation involved may not differ from the typical pattern of variation within a species. It is natural that geneticists should favour this view, since, if it is true, it means that no processes are involved other than those which are amenable to the usual types of genetic analysis. There is also a good deal of justification for this so-called neo-Darwinian view, since, as has been explained in earlier chapters, it has been possible to demonstrate within single species all those kinds of differences which separate species. Yet some biologists have challenged this standpoint. Of these, the most convincing is Goldschmidt. Although his views have been criticized by most recent writers on evolution, they should not be rejected without a hearing. A discussion of Goldschmidt's views will bring out some of the difficulties of evolution theory.

Goldschmidt is unconvinced by arguments of the kind presented in this book tending to show that geographical races may evolve into species. He thinks that the difference between two species inhabiting the same region, and either unable to interbreed or producing sterile hybrids if they do cross, is of a different order from the difference between two geographical races, and that the latter type of difference could not evolve into the former. Since a large part of this book has been devoted to presenting the opposite view, there is no point in restating the case here; however, the examples of gulls and of tits given in Chapter 13, and of the Galapagos finches described in Chapter 14, would be difficult to explain in any other way than by assuming that geographical races have in fact evolved into species. In rejecting the orthodox view, Goldschmidt has been forced to seek for some other process which can account for the origin of species and of larger groups. This process, he suggests, is 'systemic' mutation leading in one or a few steps to large phenotypic changes.

The difficulty in accepting the evolutionary relevance of 'large' mutations is in understanding how a large change could also be an adaptive one. The origin in one step of a series of independent adaptations to a new way of life would not be a mutation, it would be a miracle. But the development of animals is such that a single major change may be compensated for by modifications elsewhere. For example, Slijper described the skeleton and muscular system of a goat born without fore legs, which learnt to hop actively like a kangaroo, He found that compared to a normal goat, there were changes in the relative sizes of muscles and ligaments, and that many bones, particularly in the vertebral column, were altered in shape. The skeleton of this goat differs from that of a normal goat as much as the skeleton of a grey lethal mouse differs from that of a normal mouse; the distinction is that in the case of the bipedal goat, all the characteristics of its skeleton are adaptive to its peculiar gait. These secondary modifications occurred because muscles which are used grow bigger, tendons grow along lines of tension, bone grows along lines of compression, and so on. The relevance of such developmental flexibility is that a single major change – for example the loss of the fore legs – instead of being a disaster may be compatible with life.

This kind of reasoning has led Goldschmidt to suggest the occurrence in evolution of 'hopeful monsters', that is, of individuals departing more or less sharply from the phenotype normal to their species, able to survive in conditions or to perform functions impossible to the rest of the species, but dependent on further selection to perfect their adaptations. A visit to any genetical laboratory will show that plenty of 'monsters' are born; if they are to be hopeful, their new organs or structures must be sufficiently well integrated with the old to enable them to survive and breed until selection has had time to adapt them further. Now, whereas random change in one part of a machine would almost certainly mean that the machine as a whole would not fit together, a random change in one part of an organism (or strictly, in one of the developmental processes of that organism) may well be compensated for by adaptive changes in other parts. It is because of this fact that a monster may perhaps be a hopeful one.

One example of a kind of evolutionary change which might take place suddenly by one or a few mutations (although it need not necessarily take place in this way) is neoteny. Another group of facts quoted by Goldschmidt in support of his views concerns the segmentation of animals. The bodies of arthropods (myriapods, crustacea, insects, spiders) are built of a series of segments, each of which bears a pair of appendages which may be used for walking, swimming, feeding, or for a variety of other functions. It is believed that in the primitive members of these groups the appendages borne on different segments resembled one another more closely than they do in advanced forms, in which the appendages of different segments are specialized for different jobs. Now a number of mutants are known in *Drosophila* which cause the appendages on a given segment to develop in a manner appropriate to another segment. Examples include 'aristapedia', in which the antennae develop as leg-like structures, and 'tetraptera' in which, in place of the club-shaped sense organs known as halteres, normally present on the hindmost thoracic segment, there develops a second pair of wings.

There is nothing miraculous about such mutations. Embryonic tissue of *Drosophila* is capable of developing into a wing, a leg or an antenna according to the influences brought to bear on it. If, for example, the embryonic rudiments of the antennae are slightly delayed in their appearance, they may come under influences inducing them to differentiate as legs instead of as antennae. In the same way the tissues of the flank of a newt are capable of developing into a leg, and can be induced to do so by a variety of stimuli, so that if, for example, a nasal rudiment is grafted into the flank of a developing newt, this may induce the appearance of an additional limb. It would, however, be a miracle if it proved possible by such a simple operation to induce a fish to develop a leg, just as it would be a miracle if a single mutation caused a centipede or a spider to develop a wing. The relevance of mutants such as aristapedia or tetraptera, therefore, is not that they suggest that wings or legs may have arisen in a more or less perfect form by a single mutation, but that major rearrangements of already evolved appendages could so arise; it does not follow that they have in fact done so.

One further example will show that the concept of 'hopeful monsters' must be used with a great caution. Goldschmidt wrote: 'A Manx cat with a hereditary concrescence of the tail vertebrae ... is just a monster. But a mutant of *Archaeopteryx* producing the same monstrosity was a hopeful monster because the resulting fanlike arrangement of the tail feathers was a great improvement in the mechanics of flying.' Now I believe it can be shown that the long tail of *Archaeopteryx*, with its bordering feathers, was not merely a survival of its reptilian ancestry (although certainly it was that), but was also a necessary adaptation to its mode of flight. Primitive flying animals, whether insects, birds, or the ancestors of the pterodactyls, possessed long stiff tails used as stabilizers, analogous to the feathers on the back of a dart or arrow. A modern bird does not use its tail as a stabilizer but as an accessory lifting surface for slow flight, analogous to the flaps of an aeroplane. This change in the function of the tail requires the evolution of sensory mechanisms and of behavioural reflexes to enable a bird to fly without a stabilizer. If I am right, then an *Archaeopteryx* with a fan-shaped tail would have been a flightless monster; the shortening of the tail was probably a gradual process, and even if it occurred suddenly, it had to await the evolution of sense organs and a nervous system capable of unstable flight. I cannot imagine the latter process occurring otherwise than gradually, by the selection of numerous modifications.

It remains true that the decisive step in the origin of new groups of animals or plants is the exposure of populations to new kinds of selection, whether the cause be a change in genetic material, in learned habits or in environment, or some combination of these factors. Further, those individuals which can best adapt, during their lifetime, to the new circumstances are most likely to be the starting-point of new evolutionary trends, because they will leave most progeny. The consequences of this fact have been discussed by another biologist, Waddington, who, like Goldschmidt, has been concerned both in genetical and in embryological research.

Waddington's ideas can best be introduced by discussing a particular problem, that of the development of calluses on the skin of vertebrates. It is a property of the skin both of birds and mammals that it should develop horny thickenings in response to pressure; it is for this reason that the sons of toil have horny hands. Yet in some cases the skin in particular regions thickens in the embryo before any pressure has been applied. This is so for the skin on the soles of the feet in man, and also of patches of skin on the rumps of ostriches, which squat on the ground. It would be convenient if this could be explained by saying that ancestral ostriches developed calluses on their rumps as a direct response to pressure when they squatted, and that the fact that this character was individually acquired in a number of successive generations caused changes in the hereditary material, or mutations, so that their descendants possess genetically determined calluses appearing before pressure is applied. But it is difficult to accept this 'Lamarckian' explanation because we know of no hereditary mechanism whereby acquired characters can be inherited in this way. The alternative, to suppose that a random mutation was responsible for the development of a callus in just the right place, seems rather implausible. Waddington has suggested a way out of this impasse in terms of two processes which he has called 'canalization of development' and 'genetic assimilation'.

Waddington's main thesis is that 'developmental reactions, *as they occur in organisms submitted to natural selection*, are in general canalized. That is to say, they are adjusted so as to bring about one definite end result regardless of minor variations in conditions during the course of the reaction.' The truth of this statement is demonstrated by the extraordinary uniformity of populations of wild animals, a uniformity which requires that development be regulated. Further, cells or organisms of a given genotype may be able to develop in any one of a number of sharply defined ways, depending on circumstances. Thus all the cells of the body are similar in genotype, yet develop as nerve cells, muscle cells, bone cells, and so on, but not as intermediates between these types. Similarly, individuals of a given species may develop as males or as females, but very rarely as intersexes. It is as if development could flow along one of a number of well-defined 'canals'. Now this canalization is a consequence of natural selection; one way of demonstrating this is to examine extreme mutant forms in laboratory populations. For example, in a stock of *Drosophila subobscura* homozygous for the recessive mutant 'aristapedia' (see page 319), some flies have antennae which differ from the wild type only in a slight thickening at the base of the hair-like 'arista', in others the whole antenna is deformed into a leg-like structure, and many intermediate conditions occur. In contrast, it is difficult to detect any differences between the antennae of flies in a wild-type stock not carrying the mutant. The explanation is that the mutant forms have not been exposed to natural selection, and consequently their development is not canalized.

What is the relevance of all this to the development of skin calluses? Waddington argues that when ostriches first started squatting, calluses developed as a direct response to pressure. But it would be a selective advantage that the size, thickness, and position of these calluses should be uniform, and not dependent on the strength of the stimulus; there is presumably an optimal kind of callus to have. Consequently, over a long period of time and under the influence of selection, the developmental processes would be modified so that a callus of uniform character should appear in response to a wide range of stimuli; that is, its development would be canalized. Now it is known that once a process is canalized in this way, it may be set going by a variety of quite simple stimuli (compare the development of additional limbs in newts described above). It would therefore be quite understandable that a random mutation should take over the role of a stimulus causing the appearance of a callus, particularly since, to quote Waddington, 'it is an advantage to the young ostrich going out into the hard world to have adequate callosities even if it were reared in a particularly soft and cosy nest'. This latter process Waddington has called 'genetic assimilation'. The combined effects of adaptation during development to environmental stimuli, canalization of development and genetic assimilation are to mimic Lamarckian inheritance without involving any process not known to occur.

Waddington has since demonstrated the genetic assimilation of an acquired character in the laboratory, using *Drosophila melanogaster*. He started with a stock of flies in which the wing veins were normal, and found that if the pupae were subjected to a heat shock at a particular stage in development, a small proportion of the flies which emerged lacked the posterior crossvein. By breeding only from such flies in successive generations, the proportion of crossveinless flies was greatly increased, and, after fourteen generations of such selection, a few flies were found lacking the crossvein although they had not been subjected to a heat shock. By breeding only from these it was possible to establish a stock of flies most of which lacked a crossvein although no heat shock was applied. In other words, a character, the absence of a crossvein, which first appeared in a few individuals in response to an environmental stimulus, was genetically assimilated so that it appeared in most individuals without the need for such a stimulus.

This is an elegant and illuminating experiment, but there are two respects in which it does not reproduce the processes which Waddington had suggested might have occurred in the evolution of skin calluses. First, there is no evidence that canalization was involved, since in the final population the appearance of the crossveinless flies was highly variable, the crossvein sometimes being wholly absent, but more often showing gaps of greater or less extent; in this respect the flies resembled uncanalized mutant forms rather than wild-type flies. Second, the response to the environmental stimulus was not adaptive to the stimulus which evoked it; there is no evidence that a fly which lacks a crossvein is therefore better able to withstand heat shocks, whereas an ostrich which develops callosities is better able to withstand pressure.

In fact this experiment is probably best interpreted in terms of thresholds, as indicated in Figure 36. Whether or not a

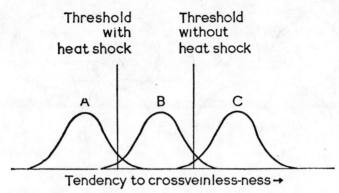

Figure 36. A possible interpretation of Waddington's experiment in 'genetic assimilation'. For explanation see text.

particular fly is crossveinless depends in part on its genetically determined tendency to develop this character, a tendency which can be thought of as a continuously varying one similar to size or bristle number. Only if this tendency reaches a particular threshold will the crossveinless phenotype appear; this threshold is lowered if a heat shock is applied. The effect of selection was to increase the tendency to crossveinlessness from condition A in the original population to condition B, in which a few flies lacked the crossvein without being given a heat shock, and finally to condition C in which most flies were crossveinless. The only relevance of the heat shock is that, in the absence of such a shock, all flies in the original population would have had crossveins, and so there would have been no way of telling which individuals had the greatest tendency to crossveinlessness, and so no way of selecting suitable parents for the next generation; the heat shock acted as a kind of indicator to show up genetic differences which could not otherwise be detected.

The relevance of this experiment to evolutionary theories seems to me to be as follows. If animals are exposed to new environmental conditions (in this case heat shock during pupation), differences will appear between them which were not apparent in the old conditions. In nature, but probably not in this experiment, these differences will usually reflect differences in the capacities of individuals to adapt to the new

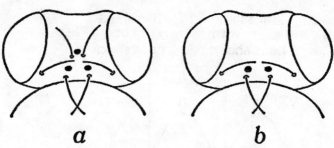

Figure 37. a. The pattern of bristles and ocelli on the top of the head of
Drosophila; b. a new pattern canalized by selection.

circumstances. Because differences appear between individuals
which otherwise would have been alike, selection can act to
change the population and to adapt it to the new circumstances.
Thus the relatively rapid evolutionary changes during the
origin of new groups are to be explained partly by changes in
selection pressure when a group is exposed to new conditions,
and partly by the appearance of new variation caused by the
new circumstances, particularly in the adaptability of different
individuals.

It is essential to Waddington's argument that development
be canalized so as to bring about a definite end result, and that
this canalization be a result of natural selection. If so, it should
be possible to choose some phenotype which occurs only
sporadically, and by selection produce a population in which
this phenotype is the normal end result of development. This
possibility has been investigated in the case of the pattern of
ocelli (simple eyes) and bristles on the top of the head in
Drosophila subobscura.

The normal pattern, found in almost every member of this
and other species of *Drosophila*, is shown in Figure 37a. A
recessive mutation, *ocelliless*, affects these structures. In the
original *ocelliless* population, over half the flies lacked all nine
structures, and the rest had one or more of them, different flies
having different structures. By breeding from those flies with the
largest number of structures, a population was obtained, still
homozygous for the mutant *ocelliless*, in which most flies had the
wild-type pattern of nine structures; it is quite common to be

able to select 'modifying' genes which mask the effects of a major mutation in this way.

But in this population, development had merely been restored to a pre-existing channel. In a second *ocelliless* population, flies having the two posterior ocelli but lacking the anterior one were chosen as parents. After fifteen generations, a population was obtained in which almost two-thirds of the flies had the phenotype shown in Figure 37b. Doubtless if selection had been continued for longer, this phenotype would have become all but universal.

It has been possible to deduce the nature of the change which has been responsible for the development of the new pattern. The head is covered by a layer of cells, the hypodermis, and it is known that each bristle develops from a single modified hypodermal cell. If a normal pattern is to develop, two things are required. First there must be some 'instructions' (called by Stern a 'prepattern') indicating which particular hypodermal cells are to differentiate into bristles. Such a prepattern can be pictured as a varying concentration of some chemical substance which, if it is high enough, can cause cells to differentiate into bristles or ocelli. Second, the cells must be 'competent' to respond to the prepattern by differentiating.

It is supposed that the prepattern remained the same in all the populations described. The original *ocelliless* population failed to form structures because the cells were not competent to respond; perhaps they lacked some 'precursor' substance essential for the formation of bristles and ocelli.

In the population in which the wild-type pattern was restored, the amount of the precursor substance was increased by selection of modifying genes up to the normal level. But in the population whose typical phenotype is shown in Figure 37b, selection increased the amount of precursor at the back of the head but removed it from the front. That the prepattern for an anterior ocellus was still present in this population is shown by the fact that an occasional fly did develop an anterior ocellus in the typical position but greatly reduced in size.

Thus selection canalized a new phenotype by creating an antero-posterior gradient of competence on the head. It is interesting that a comparable experiment, breeding always from flies having the left ocellus but lacking the right one, had no effect; it proved impossible to create by selection a left-right gradient. This is because there is in any case a difference between the front and the back of the developing head of a fly, which can form the basis for the later appearance of an antero-posterior gradient of competence, but there is no comparable difference between the left and right sides of a developing insect. This illustrates a point made earlier (page 65); the kind of new heritable variation which can appear in a population, and hence its possible directions of evolution, depend on its present mode 'of development. Which of these possible directions evolution will in fact take depends on natural selection.

After a discussion of the fossil record and the origin of major groups, the changes which occurred in the experiments described in this chapter seem inadequately small. This apparent triviality is an inevitable feature of evolutionary changes occurring in the laboratory in a few tens of generations. But they do illustrate the point that an understanding of evolution requires an understanding of development.

CHAPTER 19

Evolution and History

About 400 million years ago the first aquatic vertebrates evolved; at least two million years ago man's ancestors first chipped stones to make simple tools. Less than ten thousand years ago, in the neolithic revolution, animals and plants were first domesticated. If a film, greatly speeded up, were to be made of vertebrate evolution, to run for a total of two hours, tool-making man would appear only in the last minute. If another two-hour film were made of the history of tool-making man, the domestication of animals and plants would be shown only during the last half minute, and the period between the invention of the steam engine and the discovery of atomic energy would be only one second.

These figures show how rapid are historical changes when compared to evolutionary ones. Even if, as is almost certainly the case, the rate of human evolution today is, as evolutionary rates go, extremely rapid, it is still slow on an historical time scale. It follows that, to a first approximation, there is no need to take into account evolutionary changes when analysing the causes of historical change.

The phrase 'You can't change human nature' is repeated more often than it is understood. There is a sense in which geneticists would regard it as at least approximately true, although not in the sense in which it is usually meant. If by 'human nature' is meant the 'nature' of individual human beings, that is to say, their genetically determined capacity, not to develop into some one particular kind of person, but to develop in any one of a variety of ways according to the circumstances of their upbringing, then the statement is true.

We do not know how to alter the genetic constitution of the human race so as to change the capacities of the individuals born, although we can say that the indiscriminate scattering of radioactive substances into the atmosphere will increase slightly but significantly the number of the genetically handicapped, and that the discouragement of marriages between close relatives will decrease that number. In fact, we can be fairly confident that the 'nature' (i.e. the genetically determined capacities) of human beings has not greatly changed since the neolithic revolution, since 7,000 years is too short a period for major evolutionary changes. There are probably genetically determined differences of a statistical kind in temperament and talents as well as in physical type between human races, and the recent increase in intermarriage between human races must have resulted in changes in the genetic constitution of the population, although we cannot at present say whether the result has been an increase or a decrease in health, fertility, or intelligence.

But if by 'human nature' is meant the kind of characters, temperaments, beliefs, prejudices, consciences, and talents which individuals in fact develop, the statement is manifestly untrue. That rapid historical changes can take place in, for example, the scientific attainments, religious beliefs, or social customs of a people is too well known to require exemplifying. But, even when attempts have been made to find more fundamental psychological characteristics, common to all human beings, it still seems likely that these characteristics are individually conditioned, and are common only to those individuals who also have something in common in the social conditions in which they grew up.

It follows that human nature consists not of some one fixed pattern of behaviour, but of the capacity to develop a variety of different patterns of behaviour in different circumstances. For example, the differences between the customs and beliefs of present-day Englishmen, of the Aztecs of Mexico, and of the aborigines of Australia do not necessarily reflect genetic differences between these peoples; although such genetic differences exist, they are probably not responsible for the cultural differences.

It has been one of the main themes of this book that animals and plants can adapt as individuals to changed conditions, and that their ability to do so is of evolutionary importance. But such adaptations are not transmitted to their offspring, although it was argued that changes which originate as individual adaptations may, after many generations of selection, be genetically assimilated. What is characteristic of man is that this capacity of individuals to adapt has been so increased that it has led to a qualitatively new process, that of continuous historical change.

The nervous systems of animals originated to make possible rapid and appropriate responses to immediate circumstances; a flatworm moves away from a strong light but towards a source of food. In higher animals, with more complex nervous systems, two rather different kinds of elaboration have taken place; sometimes, as in birds, both types of elaborate behaviour have reached a high stage of evolution in the same species. On the one hand, complex series of instinctive acts are performed in response to specific stimuli, without any need to learn either the relevance of the stimulus or the effectiveness of the response. On the other, individuals can store past information, and can respond in the way which such past experience suggests will be appropriate. These two kinds of behaviour, 'instinctive' and 'learnt', are not sharply distinct, and both occur side by side in all higher animals. But in man the instinctive component is difficult to recognize; there is no pattern of behaviour more complicated than the sucking of a baby which does not require to be learnt. In contrast, man's capacity to store information, and to use it to ensure that his future actions will be appropriate, is enormously greater than that of any other animal. This superiority cannot by itself account for the rapidity of historical change. Men are also able to communicate their experience by speech, and later by writing, to their fellows and to subsequent generations. There is therefore no need to wait for the genetic assimilation of a new adaptive advance made by an individual. Such advances are transmitted to future generations by cultural and not by genetic means; children do not inherit a 'racial memory', they learn what their parents teach them. Finally, men make tools, and so can change their environment to suit themselves instead of evolving new, genetically determined adaptations to new environments.

It is this change from genetic to cultural transmission which determines the differences between evolutionary and historical processes, and is responsible for the greater rapidity of the latter. But there is no one time in the past when evolution stopped and history started. The recent spread among tits of the habit of removing the tops of milk bottles was an historical change, as was the invasion of Australia by rabbits; neither event depended on genetic changes in the populations concerned, i.e. on evolutionary changes. Historical events of this kind must have occurred throughout geological time, but they have been intermittent and sporadic. Instead of one historical event leading directly to another, as in human history, among animals each such event has been followed by long periods of genetic evolution.

Similarly, evolutionary changes did not stop at the dawn of human history. The tragic extinction of the natives of Tierra del Fuego, many of whom died in measles epidemics which the European immigrants survived, was in part an evolutionary event, since the survival of one group of human beings rather than another depended on a genetic difference between them. It was also in part an historical event; the Europeans had rifles. Of all the changes in the selection pressures acting on man since he lived in large communities, probably the most important has been the greater importance of infectious diseases. Although there is no direct evidence, it may be that the most significant genetic difference between ourselves and our ancestors of 10,000 years ago lies in our greater resistance to such diseases.

Despite this overlap in time between the two processes, in studying the origin of man we are concerned with a period in which historical processes gradually replaced evolutionary ones as the main causes of change. Men made and used stone tools for some two million years before the domestication of animals and plants in the neolithic revolution. The latter revolution was accompanied by a whole series of other technical advances, including polished stone tools, pottery, the wheel (for transport, pottery, and spinning), and writing. From that time until the end of the Middle Ages, the rate of technical advance was high compared to that during the palaeolithic, but still uneven and spasmodic. For the last few hundred years advance has been continuous and extremely rapid.

Does this acceleration in the rate of technical progress reflect an evolutionary increase in human intelligence? Or does it merely reflect the fact that the first steps were difficult to make, and later developments relatively easy? Both explanations are partly true. The early difficulties are of two kinds, one intrinsic to the problem of invention itself, and the other arising from the limited opportunities and unscientific frame of mind inevitable in the members of a primitive society. The intrinsic difficulties of the first inventions are obvious. All that the modern inventor has to do, in most cases, is to combine in a new way inventions and discoveries already made. For example, the bicycle was a most ingenious invention, yet a knowledge of the wheel, of the smelting and working of metals, of the pedal, and of the chain and sprocket drive were all necessary preconditions, as was the existence of surfaced roads if the invention was to be of any use. Yet for primitive man who had mastered few techniques, little progress could be made by 'hybridizing' existing inventions, and advance depended on more completely original discoveries and inventions, such as the wheel and axle, or the making of fire.

The frame of mind of early men would also tend to hamper technical progress, since they were probably more concerned when making a tool to reproduce exactly a pattern handed down to them by their ancestors than to try out new ways of doing things. For example, 'hand axes' of the Acheulian culture were made for a period of over 100,000 years, over much of Africa, western Europe, and southern Asia, from a variety of types of stone. These axes were some nine inches long, with a rounded and a more pointed end, flat on one surface and convex on the other, and with the edges sharpened by removing flakes from the central core which formed the tool. This wide distribution in space and time of a particular type of tool implies a process of meticulous copying for many generations. And after all, why not? The hand axe, although not an axe, was presumably an effective general purpose tool, and certainly far more useful than most of the tools which would have been produced by a more experimentally minded flint-knapper.

In fact, the value of the policy of 'try it and see', which lies at the root of modern technical advance, was not recognized till the Renaissance. Although many earlier advances must have been made because an individual performed what today we should call an experiment, the appreciation that the experimental method is one of general application is quite recent. It probably owes its origin to the emergence, after the Middle Ages, of a class of men sufficiently leisured and sufficiently literate to be capable of formulating general ideas yet, because they no longer held the prejudices against practical activity which are almost inevitable in the members of a slave-owning society, not ashamed to be interested in the processes of production.

The technical advances during the neolithic revolution probably depended less on changes in outlook than on changes in economic circumstances. An agricultural society could provide a sufficient surplus of food to support a certain number of full-time artisans, and also of people concerned with the organization of irrigation and of the storing of food, and it was to these new classes that much of the progress was due. Yet the domestication of animals and plants itself required a certain level of technique as a precondition; for example, the growing of cereals requires methods of breaking up the ground before sowing, and also of reaping, threshing, grinding, and storing the crop. Thus domestication was necessarily preceded by a period during which other techniques were gradually acquired. Progress was by later standards slow because of the intrinsic difficulty of the first inventions and discoveries, because of the lack of leisure in a food-gathering community, and because of conditions which favoured tradition rather than experiment.

There is therefore no need to assume that the intelligence of palaeolithic men was lower than our own, in order to explain the recent acceleration in the rate of technical progress, since a plausible explanation of the facts can be given in historical terms. But it does not follow that this is the whole story. It may be that evolutionary changes in intelligence and historical advances in technique proceeded side by side during the palaeolithic. To decide this question, it would be desirable to compare the types of tools found throughout this period with the physical type of the men who made and used them. It is, of course, impossible to deduce a man's intelligence from his skeleton, since brain size as measured by cranial capacity is only an extremely rough guide to intelligence. But at least a change in skeletal type indicates that evolution has taken place, which may also have affected intellectual capacity. There are other difficulties. The only tools which remain are of stone (and, towards the end of the palaeolithic, of bone and antler) although probably the commonest artefacts were of wood, and have left no trace. Yet an Australian aborigine, using an unflaked but naturally sharp stone, can fashion an effective wooden spear-thrower.

Even more serious is the incomplete nature of the fossil record. It is rare to find a complete skeleton; more often we have to be content with a fragment of jaw bone and a few teeth. A particular site, such as Chou Kou Tien or Swartkrans, may provide evidence about the kind of men living in a particular place at a particular time, but for most times and for most places known from the evidence of tools to have been inhabited we have no information. Our picture of human evolution is therefore conjectural, and likely to change in important ways as more information becomes available.

The first fossils thought to lie on the human rather than the anthropoid ape line of descent have been named *Ramapithecus*. They are of Miocene age, some ten million years old. Unfortunately, all we have are fragments of jaws and teeth. They come from a small primate, perhaps three feet tall. The tooth row is smoothly rounded, and the canines are small. In this they resemble the teeth of man and not of the great apes with their greatly enlarged canines and elongated and rectangular palate. This is one of the main anatomical differences between man and the apes. It may have an important ecological significance. Canines are important to primates not in killing their food but as a defence against predators. If *Ramapithecus* had lost its canines, this suggests either that it lacked predators, which seems unlikely, or that it had evolved some other means of defence. Chimpanzees will use sticks, either as clubs or missiles, against threatening predators, although with little skill or effectiveness. The suggestion is that *Ramapithecus* was already making effective use of weapons in defence, and hence could afford to be without canines. If so, *Ramapithecus* was already committed to the human road as a user of tools held in the hands. But this is speculation, which can only be confirmed when more of the skeleton has been found.

The next stage of human evolution is represented by the Australopithecines of South and East Africa. Much more is known of these early men; complete skulls and most of the postcranial skeleton have been found. They stood some four feet high, and walked erect, as can be deduced both from the pelvis and leg bones, and from the position of the foramen magnum, whereby the spinal cord enters the skull; the latter shows that the skull was balanced on top of the backbone rather than projecting forward from it. The pelvis, however, is not identical to that of modern man, and the gait was probably not fully human. The brain size is small, being approximately 500 c.c. in comparison with the range of 1,300–1,500 c.c. typical of modern man. The sites are not always easy to date, but it seems that they lived in Africa from at least five million years ago to as recently as one million years ago.

Australopithecus is now known to have manufactured simple stone tools. At Olduvai in East Africa, Louis Leakey found a living-floor where tools had been manufactured from stones which had been transported a considerable distance to the site. There is also strong evidence that they used bone tools, and doubtless they also used wood. They did not possess fire. Their food consisted at least in part of animal game. Antelopes of various sizes provide the commonest bones found in association with their living areas, but it seems that they also occasionally fed on larger game, including wild cattle, giraffes and rhinoceros. It does not follow that they were able to kill these larger mammals, since they may have been scavengers after lions and other carnivores. An odd feature is the large number of baboon skulls found in caves occupied by *Australopithecus*. A majority of the skulls have depressed fractures. Dart has argued that these fractures were caused by a blow from a club, probably the femur of an antelope, and that the blows were usually struck from the front by a right-handed assailant. Other investigators believe that these baboons were killed by leopards.

Australopithecus is thought to have lived in open savannah country. It is not clear whether their range extended outside Africa. As we shall see later, the known African populations may be too late in time to be our ancestors. There is one remaining puzzle to be discussed. Was there one species of *Australopithecus* in Africa, or did several species coexist there? Skeletons of very different physical types are known. There is in particular a striking contrast between a 'robust' and a 'gracile' type. There is still controversy as to whether these represent the males and females of a single species, or two distinct species. If they are one species, then the sex difference in size and strength was much greater than in *Homo sapiens*, but it is not outside the range found in other primates. The best evidence for regarding them as distinct species is that the two types are sometimes found in different places or at different times.

The next group of hominids to be discussed can conveniently be called *Homo erectus*, or 'pithecanthropines'; the latter name arises because the best known specimens (Java man and Peking man) were once named *Pithecanthropus erectus*. Java man dates from 800,000 years ago, and Peking man is perhaps half that age. They are similar in having long low skulls with massive brow ridges, and with a cranial capacity of the order of 1,000 c.c. – twice that of a typical *Australopithecus* but substantially smaller than modern man. As the name *erectus* suggests, their pelvis and leg structure is indistinguishable from modern man. Our knowledge of their way of life comes mainly from the caves at Chou Kou Tien. They had fire, and manufactured a more varied set of stone, bone and antler tools than *Australopithecus*. They successfully hunted big game, but one can only guess at how they succeeded in capturing bison, horses or rhinoceros; one possibility is that they drove the game towards a natural trap such as a bog or cliff. One startling feature is that many of the *H. erectus* skulls have been broken open from the base surrounding the foramen magnum; a possible explanation is that they practised ritual cannibalism.

Outside China and Java, pithecanthropine remains are few and far between. From the few skull fragments which exist, it is thought that the *H. erectus* fragments resembling those of China extended right across Asia and the Middle East into Africa. Particularly interesting are fragments of skull from Hazorea in Israel, tentatively identified as *H. erectus*, found in association with an Acheulian hand axe industry; *H. erectus* remains have also been found associated with Acheulian hand axes at Olduvai in East Africa. Despite the rarity of fossils, the evidence from stone tools makes it clear that half a million years ago some form or forms of men lived in the tropical and temperate regions of the old world from China to the Mediterranean.

In the last few years, two finds have been made which suggest that men of this type are far more ancient than had previously been thought. First, the oldest of the beds in Java in which fossil *H. erectus* have been found, the Djetis beds, have been dated by the potassium-argon method, and found to be 1·8 million years old, more than twice as old as the later fossils from the same area. Still more surprising, Richard Leakey has found a skull and other bones at Lake Rudolf in East Africa dating from three million years ago which, from skull structure, cranial capacity and length of limb bones, he has interpreted as a form of *H. erectus*. This means that *H. erectus* may have coexisted, not necessarily in the same habitats, with *Australopithecus* for over two million years. This is a puzzling and unexpected conclusion.

There are few fossils to help us in tracing the evolution of modern man, *H. sapiens*, from *H. erectus*. What evidence there is suggests that there was a gradual evolution from the condition found in Peking man to that of modern man, and that an effectively modern skull structure evolved rather early, more than a quarter of a million years ago. One piece of evidence for the early evolution of effectively modern man is a skull found at Swanscombe in Kent, associated with an Acheulian culture, and approximately half a million years old. Only the back part of the skull is known, but it is, at least for those parts which have been preserved, indistinguishable from that of many modern men. The cranial capacity is approximately 1,300 c.c. It is interesting that the Acheulian culture has been found associated with skulls classified, albeit tentatively, as *H. erectus* and *H. sapiens*.

The subsequent evolution of man, however, has one extremely puzzling feature. Towards the end of the last interglacial period, about 100,000 years ago, and during much of the last Ice Age the predominant culture in Europe and North Africa was the Mousterian. The people of this culture made stone tools from flakes, struck from a central core, and then further sharpened by removing minute flakes by applying pressure to the cutting edge. The stone tools made are more varied and sophisticated than those of any earlier stone culture, and for the first time stone spearheads appear. The men of this Mousterian culture belonged to a physical type sharply distinct from modern man. Their skulls have enormous brow ridges; the face and teeth are very large; the skull is long and low, but the cranial capacity is if anything larger than that of modern man. They were short and very heavily built, but the earlier belief that they walked in a stooped posture with bowed legs is now thought to be mistaken. These 'Neanderthal' men are now regarded as a race of *H. sapiens*, mainly because of their large brain size. They were certainly modern in some of their habits. They buried their dead – this is the reason why we have such relatively abundant fossil material – leaving funeral offerings of food and tools, and in one case covering the body with wild flowers.

It was once believed that Neanderthal man was ancestral to modern man, a reasonable belief in view of his apelike characteristics. Yet the recent discovery of the Swanscombe skull and of other skulls, earlier in time but more modern in appearance than Neanderthal skulls, makes this unlikely. Further, a comparison of Neanderthal skulls from early and late in the Mousterian period shows that there was a tendency for the apelike characteristics to be accentuated as time went on,

The probable explanation of these facts is as follows. From men resembling Peking man, but earlier in time, there evolved men with skulls not unlike our own, but still with marked brow ridges and somewhat prognathous jaws. Swanscombe man may well have been of this type, and an almost complete skull found at Steinheim, of early but uncertain date, answers to this description. Men of this kind were responsible for the early palaeolithic industries of Africa. From such men there were two independent lines of evolution, one culminating in modern man by the further reduction of the brow ridges and flattening of the face, and the other culminating in Neanderthal man. The selective forces responsible for the reversal of the direction of evolution in the latter line of descent are not known. Towards the end of the last ice age, some 40,000 years ago, Neanderthal man disappeared from Europe, and was replaced by men of modern type. These people possessed a greater variety of stone tools, and worked in bone, antler and ivory; their descendants are famous for their paintings in the caves of southern France.

In building up a picture of our origins, it is natural to supplement the direct fossil evidence by a study of our living relations. The primates, the mammalian order to which we belong, are a tropical and subtropical group. In the main they are arboreal, although the baboons and a few other species live on the ground in the open, taking refuge in trees or on cliffs to sleep. They are vegetarian, but feed also on insects and other small animals, only rarely killing larger prey. They are in some ways an unexpected group to have given rise to a tool-using and ultimately technological species, since they show few incipient signs of such a development. No primate species stores food, and only a few construct any kind of nest, and then only of the simplest kind. It is the lack of these two adaptations which has confined them to the tropics, leaving the temperate forests to the arboreal rodents.

We owe to our primate ancestors our grasping hands and binocular vision; although originally adaptations for climbing, these two features enabled our ancestors to manufacture and use tools. Tool use is not common among primates. Wild Capuchin monkeys use stones for breaking nuts and other purposes. In captivity, chimpanzees show considerable ingenuity in using and even in constructing tools, and they do make some use of tools in the wild; for example, they will use a leaf as a sponge to help in drinking, and a stick to help in extracting termites. Alone among primates, chimpanzees have been observed to cooperate in hunting. Jane van Lawick-Goodall observed a group of adult males hunting a Colobus monkey; one climbed into a tree to catch the monkey, while the others guarded the feet of all the trees connected at the crown to the tree containing their prey. Cooperative hunting of this kind is typical of carnivores and even of dolphins, but very unusual in primates.

We are therefore faced with something of a puzzle. Once the use of tools plays an important part in survival, there will be strong natural selection for increased intelligence. But primates seem to show a higher level of intelligence than other mammals even though few of them use tools. The high intelligence of monkeys and apes may be merely an anthropocentric judgement based on their similarity to ourselves, since there is no generally accepted way of measuring or comparing the intelligence of animals. Yet the impression of high intelligence given by monkeys and particularly by chimpanzees, is certainly a strong one. Assuming the impression is correct, why should primates be intelligent? Michael Chance has suggested that in primates intelligence has evolved as an adaptation to life in societies. In a primate society each individual must learn to behave appropriately to all the other members of the group, depending on their age, sex, and individual peculiarities. A primate can even learn to manipulate others; for example, a subordinate male Macaque monkey may approach a dominant male taking with him an infant, which will inhibit the aggression of the dominant male. What Chance is suggesting is that intelligence first evolved to cope with living in society and was later applied to the control of the material world through the use of tools. I find the argument persuasive. It may at first sight seem odd that intelligence evolved in a social context could be applied to a non-social one. To this day, however, we readily appreciate analogies between the control of society and of material objects, a fact which is deeply embedded in our language. In planning this section, I had intended to illustrate it by referring to the use of the word 'force' in physics and in politics, but I find that I can do better. I notice that earlier in this paragraph I used the word 'manipulate' to describe the use of one individual to control the behaviour of another; the root of the word is the latin *manus*, or hand. I could not want a clearer proof of the fact that we continuously borrow words and concepts from the field of social relations and apply them to material ones, and vice versa. If Chance is right, the intelligence we use today in theoretical physics first evolved to help us live in society.

Life in society may have been an important selective force bringing about increasing intelligence in primates. However, if it had not been for our grasping hands and the habit of using tools which they make possible, it is not clear why our intelligence should have evolved beyond the level found in dogs, which also live in societies.

Social life is an important ecological adaptation in many primates, and is particularly well developed in ground-living monkeys, in which it plays an important role in defence against predators. Typically, each individual belongs to a group which has a defined and stable membership and which lives in a fixed place. The size and structure of the group varies greatly from species to species. The main distinction is between species with one-male and multi-male groups. For example, in the plains-living Patas monkey, the group consists of a single adult male, several females and their young. In the Hamadryas baboon there are also stable one-male groups, but these band together into large troops for sleeping. In other baboons, there are multi-male groups, and relations between a male and female are transitory, lasting only during a single oestrus; the only long-lasting relationship is that between a mother and her young. The great apes, so far as they are known, are somewhat atypical. The gibbons form permanent monogamous pairs. The chimpanzees appear to form no permanent groups, but come together and break up again as food supply and other circumstances dictate; unlike other primates, females are not the property of a particular male, and a subordinate male can copulate with a female in the presence of a dominant male without interference.

The advantages which a species derives from social life are of two main kinds. The first is defence against predators; a large group is better at spotting and at mobbing and driving away a predator. The second is derived from the knowledge and experience of the older members of the group. Remembered knowledge of seasonal sources of food, or the location of water holes during a drought, may be crucial for survival. Knowledge of what is and what is not edible can be socially transmitted. The different patterns of society are presumably adaptations to different ways of life, but as yet the relationship between ecology and society is not well understood. Kummer has argued that the size of a group will be determined by the distribution of limiting resources. This is best understood from an example. In the Hamadryas baboon the one-male group is the largest group which can feed in a single acacia tree, the main food of this species. In contrast, the cliffs to which this species retreats at night are few and far between, but each cliff will house many animals; consequently troops of several hundred individuals are formed to utilize them.

Since primate societies are so various, it is difficult to be dogmatic about our own immediate ancestors. *Australopithecus* and later hominids supplemented food gathering with co-operative hunting of larger game. This would have required a larger group than the one-male group. Today there is almost always a division of roles between the sexes, hunting being carried out by men. Since young primates remain relatively helpless for a long time, and, unlike young carnivores, cannot be left in a safe lair, this role division may be very ancient. It should not be necessary to point out that the fact that a practice is ancient is no argument for perpetuating it; cannibalism is probably half a million years old, and may be much more. The commonest social structure for modern man is of monogamous pairs within a larger group. No comparable society is found in other primates. Perhaps the nearest approach is in the Hamadryas baboons, with one-male groups joining to form a troop. But the differences between the two social patterns seem more important than their resemblances. A male Hamadryas baboon usually 'owns' several females; he maintains ownership by constantly watching his females, and punishing a female by biting her if she wanders away or interacts with any other member of the troop. This mechanism for maintaining the pair is quite different to that found in man; since it depends on the continual presence of the male, it would be incompatible with the evolution of cooperative hunting by males.

Our own species appears to have evolved the capacity to form rather a stable pair bond between a male and female, capable of surviving periods of separation required for hunting large game. Although we do not seem to be as firmly committed to monogamy as gibbons or bullfinches, the custom is so widespread as to be almost universal among existing human societies, and it is a reasonable guess that the custom is an ancient one. The significance of monogamy as an adaptation to hunting big game has been emphasized by Desmond Morris. Although I find the argument persuasive, I am less convinced by his suggestion that other human characteristics – in particular nakedness, and the fact that the sexes are attractive to one another, and copulation can take place at all times and not only during a restricted period of oestrus, have evolved because of their role in cementing the pair bond. Other mammals and birds have evolved very stable monogamous pairs without emphasizing the importance of copulation. Nakedness, and the abundant distribution of sweat glands over the body surface, seem more likely to be a physiological adaptation to prolonged exertion in a hot climate.

At the beginning of this chapter I contrasted evolutionary and historical change. Recent studies by a group of Japanese workers of a colony of macaques on the island of Moshima have shown that cultural changes occur in primates. This wooded and mountainous island had a wild macaque population, originally confined to the forest. Since 1952 food has been put out on the beach, and this has offered new ecological opportunities. As one example of the many behavioural changes which have occurred, members of the troop now regularly wash sweet potatoes in the sea to remove sand grains before eating them. This trick was first learnt by a single young female, Imo, and gradually spread through the group, being first copied by companions of her own age, and later taught to their children;

characteristically, adults older than Imo never learnt. Today this troop has a number of behaviour patterns which were absent in 1952, and which are culturally transmitted from generation to generation; they even bathe in the sea for pleasure. Like the tits in Britain who have learnt to remove the tops of milk bottles, this is a historical and not an evolutionary event.

Perhaps the most crucial feature of all, that of language, has not yet been mentioned. Since the chimpanzee Washoe has been taught to use deaf-and-dumb sign language, it is no longer possible to assert that there is some peculiar feature of human language for ever inaccessible to animals – not that this will stop people asserting it. Washoe can use a word to describe an object which is not present. She can put words together to form meaningful 'sentences' which she has never met before; as an example of her linguistic creativity, she invented the phrase 'dirty good' to describe her pot. She is now teaching deaf-and-dumb language to younger chimps. Of course, she is using a language which was invented by human beings. Chimps, like many other animals, do communicate important information in the wild, but the range is limited. We do not know when our ancestors first used language for more abstract communication, but we may soon do so. Geschwind has recently shown that although in general it is difficult to tell much about the brain from a cast of the inside of the skull, it is possible to tell whether there was the characteristic asymmetrical expansion of the speech area. An appropriate study of fossil skulls should help to tell us how ancient speech is.

Finally, something must be said of the probable future of human evolution. Evolution tends to adapt the nature of animals and plants to their environments. In history, man has adapted his environment to his nature. In so doing, he has made it possible for many individuals to survive and to reproduce who in previous times would have been ill-adapted to do so; individuals, for example, who suffer from diabetes or, like the writer, from severe astigmatism. As we learn to recognize and to cure more hereditary diseases, this process will be accelerated. The consequence will be to increase the frequency in the population of individuals with genotypes which in the past would have rendered them unfit. In so far as advances in medical science make it possible for such individuals not only to survive but to contribute to the community, this is not a very serious matter, and the disadvantages seem trivial compared with the advantages from a humane point of view.

But it is natural to wonder whether we shall not in the future wish to exercise some control over our own evolution. The task is a very difficult one, even if we confine our attention to the negative aim of reducing the frequency of well-defined genetic incapacities caused by genes at single loci. The problem is very different, according to whether a condition is caused by a dominant or a recessive gene.

For conditions due to a single dominant gene, both the frequency of the condition and of the gene causing it would decrease if affected individuals had no children. In most cases, legislation along these lines would rightly be resented, but it may be hoped that with the spread of public understanding of genetics the desired result may be brought about by voluntary action.

For conditions due to recessive genes the problem is more difficult. The frequency of individuals homozygous for such genes, and therefore affected, has probably decreased in recent years with the decrease in frequency of marriages between close relatives. A further fall in the number of affected individuals could be brought about, for those genes for which it is possible to recognize the heterozygote, by discouraging marriages between individuals heterozygous for the same deleterious genes.

But these measures, although decreasing the number of affected individuals born, at the same time increase the frequency of the gene in the population; this happens because, if steps are taken to prevent homozygotes being born, selection cannot act to decrease the frequency of a harmful recessive gene. This will not matter for a very long time. If we can postpone the birth of individuals suffering from a genetically determined disease which is at present incurable, we may hope that when such a birth takes place in the future we may have learnt to cure the disease. But ultimately it may be necessary to take measures which will decrease the frequency of harmful recessive genes instead of merely decreasing the frequency with which individuals homozygous for them are born. It is not at present possible to see how to do this. To sterilize individuals homozygous for such recessives would be not only barbarous but ineffective. To sterilize all those individuals known to be heterozygous for at least one harmful recessive gene would, as our knowledge increases, lead us to sterilize most of the human race. But this problem is a very long-term one indeed. For the present all we need do is to extend our knowledge of human genetics, to encourage voluntary measures which will decrease the frequency with which affected individuals are born, and to refrain from those actions which will increase the mutation rate and so make the future problem more difficult. In fact it now seems quite possible that the problem will in the end be solved, not by any interference with the freedom of individuals to marry whom they wish, and to have children if they wish, but by more direct methods. One recent genetical discovery suggests that it may one day be possible to produce directed 'back mutations' at specific loci, that is to say, to change a particular harmful gene carried by an individual into its normal allele. This is the discovery of 'transduction' in bacteria, whereby one allele in a bacterium is replaced by another, the transfer being made by a virus. It will be a long time before this discovery can be turned to practical account in human genetics, but at least the theoretical possibility seems to exist; indeed the prospect is less distant than it seemed when the first edition of this book was written.

Human beings already attempt to direct the evolution of their domestic animals and plants. They have emancipated themselves from the necessity which requires all other organisms to evolve if they are radically to alter their way of life. It seems likely that they will in the future learn to control and direct their own evolution.

Further Reading

INTRODUCTION TO THE CANTO EDITION

Dawkins, R. *The Selfish Gene*, 2nd ed., Oxford University Press, 1989.
Edey, M. A. and Johanson, Donald. *Blueprints*, Oxford University Press, 1990.
Diamond, J. *The Rise and Fall of the Third Chimpanzee*, Radius, 1991.
Ridley, M. *The Problems of Evolution*, Oxford University Press, 1985.

CHAPTERS 1, 2

Darwin, C. *The Origin of Species*, Penguin Books, Harmondsworth, 1968.
Sheppard, P. M. *Natural Selection and Heredity*, 3rd ed., Hutchinson, London, 1967.

CHAPTERS 3, 4

Watson, J. D. *The Molecular Biology of the Gene*, 2nd ed., Benjamin, New York, 1970.

CHAPTER 5

Jukes, T. H. *Molecules and Evolution*, Columbia University Press, New York, 1968.

CHAPTER 6

Orgel, L. E. *The Origins of Life*, Chapman & Hall, London, 1973.
Margulis, L. *The Origin of Eukaryotic Cells*, Yale University Press, 1970.

CHAPTER 9

Lerner, I. M. *Genetic Homeostasis*, Oliver & Boyd, Edinburgh, 1954.

CHAPTER 10

Sheppard, P. M. op. cit.

346

Dobzhansky, Th. *Genetics and the Origin of Species*, 3rd ed., Columbia University Press, New York, 1951.

CHAPTER 11

Lewontin, R. C. *The Genetic Basis of Evolutionary Change*, Columbia University Press, New York, 1974.

CHAPTER 12

Williams, G. C. (ed.) *Group Selection*, Aldine-Atherton, Chicago, 1971.

CHAPTERS 13–16

Dobzhansky, Th. op. cit.
Mayr, E. *Animal Species and Evolution*, Harvard University Press, Cambridge, Mass., 1963.
Solbrig, O. T. *Principles and Methods of Plant Biosystematics*, Macmillan, London, 1970.
Stebbins, G. L. *Variation and Evolution in Plants*, Oxford University Press, Lindon, 1950.

CHAPTER 17

Simpson, G. G. *The Major Features of Evolution*, Columbia University Press, New York, 1953.

CHAPTER 18

Waddington, C. H. *The Strategy of Genes*, Allen & Unwin, London, 1957.

CHAPTER 19

Pilbeam, D. *The Evolution of Man*, Thames & Hudson, London, 1970.
Kummer, H. *Primate Societies*, Aldine-Atherton, Chicago, 1971.

Index

348

Hartley, P. H. T., 234
hawkweed, *see Hieracium*
Hennig, W., 6
hermaphroditism, 201, 206, 267
herring, 48
heterogametic sex, 254
heterokaryon, 205
heterosis, 162–3, 175–81, 186–8
heterozygote, definition of, 55
Heuts, M. J., 248
Hieracium, 140
Hiesey, W. M., 140, 143
homeobox, 22
homeostasis, 85
Homo erectus, 24, 335–7
Homo sapiens: enzyme polymorphism in,
    186, 192; origin of, 335–7
homozygote, definition of, 55
horn growth, 64
horses: evolution of, 276–8, 285–90;
    legs, 27–9, 289–90; teeth, 33,
    276–8, 288–9
Horseshoe Crab, *see Limulus*
Howard, L., 133
Hubby, J. L., 185
Huxley, J. S., 238
hybrid breakdown, 255–6
hybrid infertility, 255, 264–7, 271
hybrid inviability, 253–5, 261–3, 271,
    273
hybrids: *Crepis*, 267; crows, 137–8; dogs,
    145; *Drosophila melanogaster × D.
    simulans*, 128; *Drosophila
    pseudoobscura × D. miranda*, 271;
    *Drosophila pseudoobscura × D.
    persimilis*, 221–2; *Gossypium*, 269;
    gulls, 220; lion × tiger, 39; meiosis
    in, 264–7; Mendelian inheritance
    in, 258–61; oak, 24; pears, 35;
    pheasants, 261; *Potentilla*, 35;
    radish × cabbage, 266; *Rana*, 253–4,
    271; *Silene*, 247; swallowtail, 261;
    teal, 247; *Triturus*, 258–61, 264,
    275; *Viola*, 247; wheat, 268–9;
    wheat × rye, 254–5
hymenoptera, society formation in,
    196–7
Hyponomeuta, biological races, 240

immune reaction, 134
inbreeding, 151, 158–61
injury-feigning, 193

instinct, 329
intensity of selection: definition of, 47;
    evolution rate and, 284–5
inversions, 98, 128–30, 222; in
    Drosophila populations, 176–80
irradiation, mutations and, 80–1
isolating mechanisms, 245–56

Jacob, F., 123
Java man, 335–6
Jepsen, G. L., 295–6
Jukes, T. H., 102, 346

Kalmus, H., 135
Karn, M. N., 171
Karpechenko, G. D., 266
Keck, D. D., 140, 143
Kerkis, J., 128
Kettlewell, H. B. D., 166–7
Kimura, M., 102, 105
kin selection, 195–7
King, J. L., 102
King, T. J., 124
koala 'bear', 33
Koller, P. C., 271
Koopman, K. F., 272
Kowalesky, W., 285
Kramer, G., 49
Kummer, H., 341, 347

Lacerta sicula, 49
Lack, D., 48–50, 231–8, 250
Lactuca, seasonal isolation, 249
Lamarck, J. B., 38, 77, 80, 320
Langlet, O., 140
language, 24–5, 343
Larus, *see* gulls
Leakey, L. S. B., 335
Leakey, R., 336
Lee, A., 155
lemmings, migration of, 199
lemurs, 234
Lerner, I. M., 346
Levene, H., 182
Levins, R., 182
Lewontin, R. C., 185, 188
Limnea, inheritance of coiling, 82
Limulus, 185
linkage, *see* genetic linkage
Linnaeus, C., 41
Lockley, R. M., 228
locusts, phases in, 83

Patau, K., 128
Patterson, J. T., 252
Pavlovsky, O., 180
Pearson, K., 155
Peking man, 335–8
Penrose, L. S., 171
Peromyscus, see mouse
pheasants, hybrids, 261
phenotype, definition of, 56
Phiomia, 291
photosynthesis, 116–17
phylogeny, definition of, 310
pigeons, 37, 244
Pilbeam, D., 347
Pinus sylvestris, clines in, 140
pipits, ecological separation, 234
Pissarev, V. E., 254
Pithecanthropus, see Homo erectus
plastids, *see* chloroplasts
pneumococci, 68
polar bodies, 58–9
pollen growth, 252
pollination, 251, 270
polyandry, 210
polygyny, 17, 210
polymorphism, definition of, 135
polyploidy, 266–70
Pontania, biological races of, 241
Potentilla, hybrids, 35
Prakash, S., 188
Price, G. R., 203–4
Primula, cytoplasmic inheritance in, 84
primates, 338–9; social organization in, 339–43; tool use in, 338
Proboscidea, *see* elephants
prokaryotes, 109, 118–20, 122
proteins: coding of, 89–96; genes and, 72–5, 87–9; molecular evolution, 96–108; polymorphism, 184–92
Pseudomonas, 107–8
Psylla, biological races of, 240–1

Ramapithecus, 23, 333–4
Rana, hybrids, 253–4, 271
Rasmuson, M., 159
rates of evolution, 276–85
recapitulation, law of, 311–12, 314
recessive, definition of, 55
recombination, 60–1, 118
regulator genes, 74, 107–8, 122–4
relatedness, genetic, 4
Rendel, J. M., 171

Rensch, B., 238
replication, 67, 70–2
repressor genes, 123
reproduction: and group selection, 198; asexual, 4–5, 200–2; rate of, natural selection and, 48–52
Rhagoletis pomonella, races of, 240
rhesus factor, 134, 274
ribonucleic acid (RNA), 72, 85; test-tube evolution, 11; messenger, 92–5, 115; transfer, 92–3, 110, 115
ribosomes, 93, 110, 115, 127
ribozymes, 11
ring species, 228–9
ritualization, 209
Robb, C., 288
Robertson, A., 158, 160
robins, 140; clutch size, 52
Robinson, L., 282
rodents, competition with multituberculates, 295–7
Romer, A. S., 291, 293, 297
rotifers, 4
Rubus, variation in, 225

Salisbury, E., 49
salivary gland chromosomes, 58, 126, 128, 130, 136, 177
salmon, osmotic regulation, 33
segregation, genetic, 59–61
Selander, R. K., 190
selection: artificial, 42, 150–64; intensity of, 47, 284–5
sex determination, 62, 207, 267
sexual reproduction, origin of, 200–6
sexual selection, 14–15, 211–15
sharks, 48
Sheppard, P. M., 136, 172–4, 261, 346
sibling species, 222
sickle cell anaemia, 99, 175
Silene: isolation by habitat, 247; polymorphism in, 185
Simpson, G. G., 287, 300, 301–2, 347
skin: calluses, 33–4, 320–2; grafting, 134; of horses, 28; pigmentation, 137, 140
skull: of dog and man, 315; of elephants, 291, 293; of horses, 288–9; of multituberculates, 297; of Neanderthal man, 337; of rodents, 297; Swanscombe, 336
Slijper, E. J., 317